America's Water

America's Water

Federal Roles and Responsibilities

Peter Rogers

A Twentieth Century Fund Book
The MIT Press
Cambridge, Massachusetts
London, England

This book was set in Bembo by The Maple-Vail Book Manufacturing Group and was printed and bound in the United States of America.

Library of Congress Cataloging-in-Publication Data

Rogers, Peter P., 1937–
  America's water : federal roles and responsibilities / Peter Rogers.
    p.   cm.
  "A Twentieth Century Fund book."
  Includes bibliographical references and index.
  ISBN 0-262-18156-8
  1. Water-supply—Government policy—United States. I. Title
  HD1694.A5R64   1993
  333.91'00973—dc20   93-1740
                                                                          CIP

Contents

Foreword

The American political system reflects many of the central values of our democracy. But, at least as expressed in the structure of the federal government, few would argue that it also produces a tidy and rational organization chart. Indeed, the accommodations and approximations that are characteristic of election campaigns and separated powers usually result in multiple executive agencies and numerous congressional committees sharing jurisdiction over any given policy area. Water, water everywhere, in the Washington context, means that, when it comes to making policy, everybody has a piece of the action.

Peter Philips Rogers, Gordon McKay Professor of Environmental Engineering and Professor of City Planning at Harvard University, analyzes the causes and consequences of this muddled administrative approach to one of our most basic resources. He points out that despite the curious structure for policymaking at the national level, we have made significant strides in dealing with water problems. The United States, after all, is blessed with sufficient natural gifts and good fortune to have done remarkably well in developing its water resources. Professor Rogers points out, however, that this bounty does not mean that we should continue indefinitely with the present fragmented approach. He offers a reform agenda for national water policy that is specific on both substantive and organizational issues.

This book touches on a number of areas of long-standing interest at the Twentieth Century Fund: it continues our interest in examining the availability of natural resources, dating from J. Frederic Dewhurst's *America's Needs and Resources* (1947), Raymond Mikesell's *Nonfuel Minerals* (1987), and Pietro Nivola and Robert Crandall's ongoing work on American energy policy; it complements ongoing work

on governance issues, such as Paul Peterson's examination of federalism; and it explores environmental concerns similar to those looked
at in Marian Chertow and Reid Lifset's examination of solid waste
management.

We are grateful to Peter Rogers for his careful analysis of the
complex problems facing policymakers. He makes a strong case for
the need to reform the patchwork of state to state and state and federal
agreements that now comprise our water policy.

Richard C. Leone, President
The Twentieth Century Fund
July 1993

Preface

This book was written in the belief that there are serious shortcomings in the way we formulate policy concerning water in the United States. It is over 30 years since Maass et al. (1962) published the *Design of Water Resources Systems* and over 20 years since White's (1971) *Strategies of American Water Management* appeared. While the specific concerns have changed, the overarching issues remain the same: the role of the federal government in establishing coherent policies, the interagency conflicts and overlaps, the importance of finance, the need for information, research, and education, and the coordination of federal and state efforts. This book claims that despite the shortcomings of the policymaking process, the current condition of America's water is relatively good. How good it remains, however, will depend to a large degree on our ability to make the technical, economic, political, and bureaucratic adjustments outlined in the book.

In writing a book that covers so broad a territory of technical, economic, and political concerns, an author needs to have many helpful colleagues and mentors. In my case I would like to express my thanks to six teachers who later became my colleagues: Professors Harold A. Thomas, Jr., Robert Dorfman, Arthur Maass, Roger Revelle, Maynard Hufschmidt, and Myron Fiering. All were great role models for a graduate student and an aspiring junior colleague. Mike Fiering read and reread various drafts of this book, but he did not live to see the final outcome. He is greatly missed by his colleagues at Harvard and by the water resources community worldwide. Another source of wise counsel and help has been Dr. Charles (Henry) W. Foster of Harvard's Kennedy School of Government. More than a passive supporter, Henry helped organize a project on water policy at the Kennedy School and invited me to coauthor "Federal Water Policy: Toward an Agenda for Action" in 1988, which gave me the opportunity to explore in writing many ideas for federal water policy reform.

Over the past several years I have discussed water policy with dozens of academics, government servants, policymakers, and others involved in the water and natural resources management scene both inside and outside the country. Two persons who are extremely knowledgeable about federal water policy and acted as sounding boards for my ideas were Theodore R. Schad and Eugene Z. Stakhiv. Others, such as Daniel P. Beard, Ramesh Bhatia, John Briscoe, Henry P. Caulfield, Mahesh Chaturvedi, Kenneth D. Frederick, David Getches, Joseph Harrington, Henry D. Jacoby, Guy LeMoigne, David Moreau, Daniel A. Okun, Keith Pitman, Martin Reuss, Kyle Schilling, Harry Schwarz, David Seckler, Gerald D. Seinwill, Gilbert White, and Richard Wilson, were at various times buttonholed and forced to listen to my version of water policy needs. They provided me with insights and documents that greatly extended my understanding of the subject. Peter Lydon read and commented upon the entire document with transcontinental insight from Berkeley, California. The book benefited greatly from his comments. I appreciate the help provided by these and others and hope that this book will reflect their inputs in molding my views. In making this list of persons who were particularly helpful I am sure to have missed some; I hope they will forgive my lapse of memory.

Much of the initial research for the book was carried out while developing and teaching a course on water policy at the Kennedy School of Government from 1983 to 1988. I want to thank the students in that course over the years for their help in refining this material. During the research for this book I also received help from several Harvard undergraduates. Thanks go to Christine Hill, Susan Karls, and Rhadika DeSilva. Linda Klaamas worked on this project for three years as a research assistant and edited several sections of the book before returning to Canada to prepare to become a political decision maker herself. Her help is gratefully acknowledged. I also received secretarial and administrative help from Heidi Montieth and Susan Morrison.

I would also like to thank the Rockefeller Foundation for providing me with the opportunity to spend a quiet month at the Villa Serbelloni in Bellagio, Italy. This was at a critical stage of the development of the manuscript and I benefited greatly from the intellectual stimulation of the company of scholars assembled at the villa during my stay there in July 1989.

Most thanks are due to the Twentieth Century Fund of New York. The Fund commissioned the study, has been greatly supportive of its development over the years, and did not abandon it as the clock ran on and on. I would particularly like to thank Carol Kahn and Jason Renker for their help in editing the book and overseeing its publication.

Finally, my thanks to Rosemarie Rogers, a connoisseur of author's prefaces, who amongst other things helped edit this one, for being a relatively good sport during the time this book was written.

Peter Rogers
Harvard University
July 4, 1993

1

Water Resources and Public Policy

*Of all natural phenomena there are perhaps none which civilized man feels himself
more powerless to influence than the rain, the sun, and the wind. Yet all these are
commonly supposed by savages to be in some degree under their control.*

Frazer, 1890, p. 13

Introduction

Each year the United States uses over 500,000 gallons (2,075 tons) of
water for every man, woman, and child. This prodigious amount of
water is made available at bargain prices averaging for all uses about
19 cents per 1,000 gallons. Municipalities charge about $1 per 1,000
gallons, but self-supplied industry and agriculture often pay less than
10 cents per 1,000 gallons; bottled mineral water sells for $4,000 per
1,000 gallons. No other commodity is used with such reckless abandon as water, no other bulk commodity is demanded at such high
quality, and no other natural resource is the subject of such intense
struggles within the federal establishment and in Congress—not even
oil.

For more than 200 years, towns, states, and the federal government of the United States have been grappling with the problems of
ensuring abundant high-quality water to all citizens, supplying flood
protection, and securing commerce along interstate waterways. Over
these 200 years public goals concerning water policy have continually
shifted because the nature of the problems has changed as the country
has grown, both in size and in affluence. The water problems resulting from geographic expansion are exemplified by a comparison of
available water in the original 13 "moist" states to that available in
the more recent 13 "mostly arid" states. The original settlers had
plentiful water supplies given the size of their population and their

uses for water. The nation subsequently expanded into arid areas where finding and securing water were much more difficult and expensive and the uses to which it was put, such as hydraulic mining and irrigation, were much more demanding than those in the earlier settlements. Increasing affluence has also led to major changes in the goals. Affluence leads first to a rise in the quantities of water used by each individual and his or her support systems, and second to a change in attitude toward the environment, from a strictly utilitarian concern for short-term human survival to an appreciation of the need for long-term sustainability of the aquatic ecosystem, both to ensure future utilitarian uses and for the sake of the ecosystem itself.

This book is about water (and water quality) and the federal government's role in controlling, regulating, and supplying it. The Constitution never mentions water, and one might ask why concentrate on the federal role when it is thus constitutionally the states' responsibility. But in fact the federal government has historically been heavily involved with water issues and has defined many areas as of "federal interest" that otherwise would not have been so. In the absence of technical competence and in some cases with unclear legal mandates on the part of the state and local authorities, the federal presence was originally essential to ensure rational development of water resources. As the nation grew and the water problems became more intense, it seemed only natural that technical bureaucracies should become as large and complex as the problems they dealt with, which further strengthened the federal role in water policy. In no other area of government did the federal government invest so heavily to build up its own scientific and technical bureaucracies. But the realities of the 1990s are quite different. The competencies of the states and the localities in water issues are greatly improved, often due to active promotion by the federal agencies themselves, and most of the needed large projects have been completed. The historical justifications for very large federal manpower allocations have all but disappeared, but the large staffs remain. Several of the agencies are currently engaged in extensive soul-searching, in part devoted to looking for new or expanded missions. Now seems a good time to trim the massive federal water bureaucracy down to a size commensurate with the expected federal role and the abilities of states and localities.

A distinction needs to be made between *federal policy* and *national policy*. Federal policy is narrower in its scope and restricts itself to the federal interest in a given problem area. It does not necessarily purport to solve national problems. Indeed, because of the geographic distribution of the water problems, there is no reason to believe that there are "national" water problems. The United States is at least two countries hydrologically, split down the middle (almost exactly along the 100th meridian) from Rugby, North Dakota, to Laredo, Texas, by the 20-inch annual rainfall contour (isohyet). The water problems on either side of this line are radically different from each other and require different solutions. The 20-inch isohyet is the physiographical line distinguishing between the "moist East" able to grow rain-fed crops and the "arid West" needing irrigation to guarantee crop yields. While the water problems are quite different between the regions, they are still subject to federal law and regulation pertaining to water; hence, federal policy must be able to deal with specific regional problems and not with "national" water problems.

The Accomplishments of Federal Water Policy

It is difficult to choose unambiguous measures of the success or failure of public policy. One obvious measure is how well the policy has achieved its stated goals. In federal water policy these goals have been changing over time. The ability of policy to adapt to the changing goals, however, has been remarkable, and broadly stated, over the past 200 years federal water policy has been successful. By 1985 over $400 billion had been spent for capital investments addressing water problems; over 25,000 miles of inland waterways had been developed, 83,000 reservoirs and dams had been built, over 88,000 megawatts of hydroelectric capacity (almost the same as for nuclear power) had been installed, more than 52,000 public utilities supplied 24 billion gallons of water each day to domestic users, more than 60 million acres of land were irrigated, more than 15,000 municipal sewage treatment plants were in operation, over 60,000 water pollution control permits had been issued to industries and other point sources of pollution around the country, and the frequency of flooding on several thousand streams had been curbed. However, as an indication of how difficult it is to measure success in water policy, consider that while all of these great benefits were being achieved, 60 percent of our original 215 million acres of inland wetlands were converted to other uses, often at the urging of federal policy and agencies, and almost 50 percent of the country's 1.5 million miles of streams, as well as an unknown percentage of our groundwater, were polluted to a significant degree.

Nationwide, hundreds of thousands of individuals work in water planning, management, and control. The federal government alone has over 90,000 employees working on water problems, spread over 10 cabinet departments, two major independent agencies, and 34 smaller agencies. The states and local governments have as many as three times this number of employees, and private sector consultants and contractors have at least another 50,000 working in the field. Water is not usually considered as an "industry"; but, if it were, it would be one of the largest in the United States. It would be by far the most capital-intensive industry, and in current annual capital expenditures it would rank a close third behind electric power and petrochemicals. No other industry is so highly regulated and receives so much attention from Congress. It is an almost impossible task to enumerate all the relevant federal water legislation, regulations, and rules; just one relatively new agency, the Environmental Protection Agency (EPA), has a six-volume compendium of water regulations.

Although most of the accomplishments in planning, management, and implementation of federal and other water policy are on the input side of the ledger (capital expenditures, organizations built, manpower employed, etc.), there have also been significant outputs for all of this effort.

For the first time in its existence, the country has reversed the trend of ever-increasing water consumption. Total water use has declined since 1980 and per capita use is now less than it was in 1965. This unprecedented change, which was missed by experts as recently as 1978, means that meeting future water quantity and quality goals will be easier than expected. But, even at these reduced levels, the United States still uses more than twice as much water per capita as any other country in the world.

Second, over the past 25 years gross water pollution due to municipal and industrial discharges into the nation's waterways has also declined, although the accompanying toxic chemical contamination has not decreased as much, and non-point source pollution has become the major source of contaminants.

Third, there have been unexpected federal legislative and institutional innovations in the last two decades that have established the bulk of the legislative and regulatory groundwork to deal with water problems until well into the next century. The Safe Drinking Water Act of 1986, the Water Resources Development Acts of 1986, 1988, 1990, and 1992, and the Clean Water Act amendments of 1987 provide the basic federal policy for water supply, wastewater control, flood control, navigation, and hydropower. The cost-sharing provisions of the 1986 Water Resources Development Act and other laws have firmly established the "beneficiary pays" principle in federal water policy. The State Revolving Loan Fund of the Clean Water Act sets the framework for innovative and helpful ways of financing federally mandated wastewater improvements. The December 1988 change in rulemaking to facilitate trading of water rights within the Department of the Interior's Bureau of Reclamation, and the Reclamation Projects Authorization and Adjustment Act of 1992, establish the precedents that will enable large amounts of water to be transferred from economically dubious agriculture to more highly valued industrial, commercial, and municipal uses in the water-short regions of the West. The Reclamation Projects Act of 1992, which also reallocated large amounts of water from California's Central Valley Project for ecological protection, signals a major shift in federal water policy toward environmental protection.

Given these accomplishments, one could imagine that the problems have, by and large, been solved and that the federal government could now devote its energies and funds to more pressing social issues that threaten the social contract, such as housing and the homeless, education, access to health care, and illegal drugs. However, as this book will urge, there is still a major role to be played by the federal government in water policy, because when one set of problems is resolved, other and frequently more complex problems come into view. For example, as the gross pollution from point sources, such as municipal and industrial sewer outfalls, has been successfully reduced, water pollution from non-point sources, such as agricultural runoff and urban storm drainage, has increased. Non-point source pollution is, however, inherently much more complex to manage and expensive to control than point source pollution. Another example of a change in problems over time was the historic response to dwindling supplies of unpolluted surface water by municipalities during the 1940s, 1950s, and 1960s. Municipalities switched to groundwater, and with over 50 percent of the drinking water currently being supplied from groundwater sources—largely unregulated—the effects of promiscuous disposal of toxic chemicals into the ground and groundwater during the same time period are now a public health hazard. Cleaning up contaminated surface water is child's play in comparison to cleaning up polluted groundwater.

Furthermore, the lifestyle and expectations made possible by affluence demand ever higher levels of service and of environmental quality, and at the higher end of the scale each further increment is much more expensive to achieve than the lower-level improvements accomplished so far. We will ultimately have to come to grips with the question of "how clean is clean enough?" Can we really afford to purchase the levels of health implied by the proposed drinking water standards when many other areas of public health are neglected due to lack of funding? Finally, other larger and more complex problems are already before us, for instance the transboundary problems of acid rain and potential global warming and the ever-present risk of drought. All of these remain as unfinished business on the federal water policy agenda.

The Need for a Coherent Federal Water Policy

If federal water policies have been successful in the past, why are so many people angry at the federal government's handling of water policies today? Anger is the only word to capture the tone of a recent Western Governors Association's white paper on federal water policy coordination. In its introductory section, the paper states:

A principal characteristic of federal water policy is that policies are made in an ad hoc, decentralized manner. No agency of the executive branch or committee of Congress is responsible for keeping an eye on the "big picture." Thus, federal water policy lacks a unifying vision or even a set of guiding principles. This state of events is not appropriate in an era in which supplies are threatened by chronic drought, likely aggravated by global warming, while demand continues to grow. A host of on-the-ground problems are created by, or at least are related to, the absence of a unifying vision, including redundance of functions across programs, protracted disputes, interagency turf battles, absence of policies, and lack of finality of many water disputes. (Western Governors Association, 1989, p. 1)

Many commentators claim that the United States has not formulated federal water policy; Anderson (1983a) refers to this as the "policy drought." For example, all of the major water commissions since 1950 speak of the need for a "federal" water policy as though such a thing did not already exist, but Admiral Ben Moreell (1955), who chaired the Water Resources Panel of the second Hoover Commission, argued that the United States has a "federal" water policy by default: all the existing federal laws, agencies, and programs do actually in sum constitute "policy." This is in line with Ciriacy-Wantrup's definition: "The term water policy refers to actions of government at various levels and in various branches (legislative, judicial, executive), affecting the development and allocation of water resources" (Ciriacy-Wantrup, 1963, p. 282). Moreell was also one of the first to argue against a unitary "national" water policy due to the country's size and the radically different needs and resources across the continent, and because according to the Constitution water is under the jurisdiction of the states (subject to *minor* restrictions). He believed that each of the states should have its own water policy and that national water policy would then be the sum of the federal and state policies. Not everybody agrees with this. The National Water Alliance (NWA), an informal voluntary group of members of Congress and water management professionals, is dedicated to achieving an unified national water policy. Because national water policy is such an all-encompassing concept, and one that has such inherent conflicts, the NWA has had little success in furthering its agenda. They may be better advised to focus their efforts on one aspect of the problem, such as federal water policy—an area over which they could surely have great influence—as distinguished from national policy.

There are at least two ways of approaching policy formulation. One way is to stress what ought to be and to spell out the most desirable policy consistent with these norms. A second approach is to look at the historical situation to see what is feasible given what has been done, what is functioning now, and what is expected in the near future. In reality, all policy prescriptions are some combination of both approaches; no one would seriously propose policy purely on the basis of norms without concern for feasibility, nor vice versa. Of the last two attempts at federal water policy reform, the one in 1977 under President Carter was heavy on norms and low on feasibility, whereas that of the early 1980s under President Reagan was heavily weighted in favor of feasibility. Drawing on these experiences, this book takes a modest approach, which emphasizes feasibility and draws its norms from the history of the development of federal policy. This approach sees the challenge as developing a federal policy stance that will allow the myriad current and proposed federal activities to work together coherently without major sudden shifts. The present goal of policy reform in the water area, then, should be to achieve coherence in federal water activities with a proper relationship to the states and the nation, and not necessarily to create new policy initiatives. The thesis of this book is that the time is ripe for the United States to take advantage of the relative breathing space afforded by the considerable successes of past water policy to reform the federal water policy *process* so that the unfinished business can be dealt with in a timely and efficient manner.

Issues of state and local water policy will not be covered in this book, except where there are specific federal interests interacting with them. The issue of regions larger than individual states *is* addressed because, under the U.S. system of government, such regional water institutions and their associated policies regularly involve the federal government. It therefore behooves the federal government to be prepared with policy prescriptions in multistate arenas.

The California Drought as Demonstration

The recent six-year drought in California provides an excellent example of some of the major changes that are occurring with respect to water policy in the United States. The years from 1987 through 1991 were the five driest on record for much of California; by the beginning of 1991 urban areas were facing as much as a 50 percent shortage of water. In early February the State Water Project reduced urban water deliveries to 10 percent of normal supply (that is, a 90 percent reduction) and eliminated all agricultural deliveries. At that time the state was two-thirds of the way through the rainfall season and it was the driest year on record. How could California survive this natural disaster?

As it turned out, California has not only survived the drought but has done so at relatively minor cost or damage. The solution to the problem can best be characterized as the exercise of political will to create markets for water. Given this fundamental shift everything else fell quickly into place. In January 1991 there were calls in the media and from citizens' groups for the new governor, Pete Wilson, to declare a statewide emergency and reallocate all of the water regardless of ownership (under state law the governor has the authority to take property but has to compensate for it). Frightened by the prospect of multibillion-dollar litigation, the top officials decided instead to institute a Drought Water Bank. The bank was instructed to purchase water from farmers and then resell it to those with the most pressing needs. There was to be no coercion; all purchases and sales were to be on a voluntary basis. After much discussion it was agreed to offer $125 per acre-foot of water (approximately 50¢/1,000 gallons) to the sellers with the hope of obtaining between 750,000 and 1,000,000 acre-feet of water. The state would then sell it to whoever wanted it at $175 per acre-foot.

By the end of June 1991, the Drought Water Bank had purchased about 750,000 acre-feet of water (400,000 from fallowed farmland, 210,000 from groundwater sources, and 140,000 from surface reservoirs) under 340 separate water sales contracts. It was a surprise to many people that such large quantities of water became available so quickly. The director of California's water agency (Kennedy, 1991) claimed that the crisis made it possible to move quickly and decisively, which he believes might not have been the case in anything less than a full drought situation. Even more surprising is the fact that, by the end of 1991, more than 200,000 acre-feet of this water remained unsold in the reservoirs due to lack of purchasers.

Other commentators (Becker, Cody, and Chite, 1991) have analyzed the effect of the drought on agriculture and natural resources and concluded that California agriculture had been able to cope well with the first four years of the drought due to its flexible system of delivery from alternate sources of water. However, increasing reliance on groundwater had caused rapid declines in the water tables in some areas and could not be sustained for many more years. Fish, wildlife, and forests have been severely affected by the drought. Gleick and Nash (1991) claim that the greatest impacts have been on the environment, and that many of the ecological effects may be irreversible; for example, the size of the delta smelt *(Hypomesus transpacificus)* population has been severely impacted by the reduced flows through the delta. They believe that while the direct impact upon agriculture is likely to be only around 2 percent of the total annual agricultural income ($18 billion), substantial economic costs (an additional $3 billion over the first five years of the drought) will be borne because of decreased hydroelectric potential and loss of tourism (during the 1990–1991 season ski resorts reportedly lost about $85 million). Peabody and his collaborators (1991) took a broader look at the problem of water shortages as a permanent feature of the California water planning scene. This is of interest because instead of focusing on ways to expand water supply they explore ways to make better use of the existing water supplies; this is called *demand management*.

In a closing system, all users become increasingly interdependent. Each use of water either reduces or increases the relative supply for someone downstream, by reducing the quantity or quality of the water that is discharged. Management of the interdependence becomes a public function. Ultimately, a closing water system requires much more management than an open system. . . . The difficult part of managing a closing system is the development of mechanisms to get all users to acknowledge their interdependence and to engage them in a negotiation process that binds them to the agreements reached. (Peabody, 1991, p. 7)

California presents the best evidence of the adaptations available in modern U.S. circumstances. While the adaptations are by no means painless, California faced an extreme situation: its 1987–1992 shortfall was far beyond what could normally be expected. The drought in California is a good test of the adaptations that the rest of the country may have to face in the coming decades.

Areas of Most Pressing Need

There are a series of important water policy issues that have historically been debated at the federal level but remain to this day unfinished business. Since World War II, there have been five major federal government examinations of water resources planning, organization, and policy, in the form of special commissions or extended oversight hearings by congressional committees. Their common thread has been an effort to distance policy formulation from the front-line water implementation agencies and bring it under the auspices of some central policy-oriented agency.

Many items on this sizable agenda of unfinished business have roots going back as much as a century; they have been documented in considerable depth, and from them an abundant harvest of ideas could be gathered. This book explains why the federal role needs to be maintained, developed, and expanded in the following areas.

Financing infrastructure. For the nine categories of public works infrastructure[1] studied recently by the National Council on Public Works Improvement (1988), the country has already invested more

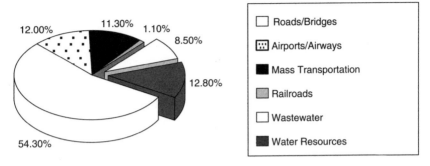

Figure 1.1
Federal spending for infrastructure, outlays by sector, 1992 (1992 total outlays = $28.2 billion). (Source: Office of Management and Budget, 1992.)

than $11 trillion in capital funds. Spending for public works infrastructure facilities in 1992 is estimated at more than $28 billion, a sum that represents approximately 6 percent of all federal nondefense outlays, which amount to $473.3 billion. But, as the council pointed out, the quality of America's infrastructure is still barely adequate to fulfill current requirements, and certainly insufficient to meet the demands of the future. The council has called for a doubling of the amount of capital the nation invests each year in new and existing public works.

Water infrastructure, much of it nearly a century old, is an important part of the national picture. Six principal types of facilities dominate this category: water supply, wastewater treatment, inland waterways and ports, hydropower, irrigation, and flood control. The sheer size of the nation's water enterprise, and its past rate of increase, are also part of the infrastructure problem. Up to 1985, the cumulative expenditures for water resources development by all parties, public and private, were estimated at more than $400 billion. From 1985 to the year 2000 an estimated additional $200 billion will be necessary to keep up with current requirements for growth and regulation. For certain aspects of water resources (e.g., navigation, flood control, hydropower, and irrigation), the federal government's share of the funds has predominated.

Despite today's budgetary problems, figure 1.1 shows that the federal government is still allocating over $6 billion annually to water-related programs and projects, which is 21.3 percent of the total annual federal expenditures on all forms of infrastructure. But it is less than one-third of the total expenditure on water by state and local

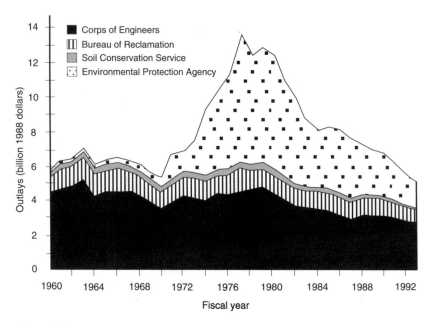

Figure 1.2
Federal outlays for water resources, 1960–1992. (Source: Office of Management and Budget, *Budget of the United States Government,* various fiscal years. Washington: Government Printing Office.)

governments and private expenditures by individuals, commerce, and industry. Figure 1.2 shows the growth and decline of expenditures on water over the past 30 years by each of the major federal agencies. Funding levels that change so radically from year to year always cause management problems in government agencies.

Although the national water resources public infrastructure appears to be in somewhat better condition than other aspects of infrastructure (National Council on Public Works Improvement, 1988), two problems loom: insufficient funds for renewal and development, and the absence of financing mechanisms for the investments required. There is also the permanent problem that must be faced in each period and each set of conditions: defining the proper role the federal government should play in building, maintaining, and financing the infrastructure improvements. Despite the large sums already spent (largely for dealing with water quantity problems), huge expenditures are still required to maintain existing infrastructure and to implement current federal policy. Merely to meet the requirements of the two most recent federal laws dealing with water quality and drinking water supply, more than $167 billion in capital expenditures will be needed between 1993 and the year 2000 (Clean Water Council, 1990). This is roughly $80 billion over and above the historical trends. Most of this expenditure will go to prevent wastewater pollution from residential, industrial, and commercial discharges. Little of this money will come from the federal government, but there will be need for federal participation, support, and management if the desired outcome is to be achieved in a balanced fashion across the country.

Privatization. The hard questions of costs, and of the sources of funds necessary to meet future water needs, remain largely unaddressed by government. To be sure, there have been encouraging steps toward objective economic analysis, and the beginnings of a process to apply uniform standards to water projects. But questions about rational water pricing—for irrigation, for commercial and industrial uses, for municipal water supply, and even for environmental quality and recreational purposes—will have to be resolved in the next decade. At a time when the federal deficit has never been higher, there is an urgent need to balance water expenditures with water revenues more closely. The question becomes crucial as the useful lifetime of many elements of the water infrastructure draws to a close. How to charge users for water to finance future water programs is likely to be the central issue of the 1990s. It is being addressed in other countries by the privatization of government responsibility for water.

In the United States there has always been significant participation by the private sector in providing water supply and water quality improvements. It may well turn out that in order to meet the financing crisis we will see much more of private water markets and private investors taking over responsibility for many urban water utilities, as in Britain and France.

Information. Decisions about water, like other aspects of societal concern, are heavily dependent upon the availability of adequate and timely information. The good news here, as the Chesapeake Research Consortium advised the Council on Environmental Quality in 1984, is that some 1,100 water-related information sources are available throughout the country, which in the aggregate go a long way toward meeting the needs of water resources decision making. However, there is little management or coordination of the knowledge. Information availability and accessibility require serious upgrading. The major substantive gap in the coverage is in information about groundwater—where the gap is huge and important, since groundwater is the source for 50 percent of publicly supplied drinking water. In short, the present system falls substantially short of meeting the information needs of the present and the foreseeable future.

Water research. Water resources research is currently diffused among at least nine federal agencies and carried out in many different manifestations and contexts. The 1992 federal budget for environmental research and development amounted to $5 billion for all departments and agencies. Less than $1.5 billion of this was allocated to federal agencies with concerns for water. In the case of one such agency, the EPA, a bare $52 million of the $502 million requested for research and development was to be spent in the program areas of water quality and drinking water. Thus, research expenditures today represent a tiny fraction of federal budgets overall, and are a far cry from the expanded and coordinated federal water resources research program called for in the 1961 Senate Select Committee report. Furthermore, the quality and emphasis of research that does go on are suffering. In its own review of the need for a national water resources research and information center, the President's Council on Environmental Quality (1984) observed that most federal research is conducted pursuant to the agencies' own mandates and missions. It concluded that the inadequacies in water research were not accidental, but rather a result of the political context in which such agencies usually operate.

Meaningful regionalism. Water is more a local and regional resource than a national resource, and water policy should, therefore, not be separated from its actual working contexts. Approaching water resources regionally by way of drainage areas or river basins, the settings in which water occurs naturally, has been the historic response to this problem. The difficulty here has been that although water occurs hydrologically in definable geographic units, within them its management is invariably subject to fragmented political jurisdictions. Efforts to deal with water geographically typically encounter strong resistance from bureaucracies that are functionally organized for different uses. While conceptually appealing, most of the attempts at water management by river basin have been failures, because, of course, U.S. society is not hydrologically organized. Human interests, efforts, and organizations tend to follow other lines—legal or political boundaries, historical economic or service areas. Property rights are a good example. Often land ownership will extend to the center line of a stream, a circumstance that virtually guarantees fragmentation of the water resource involved. In considering the best spatial arrangements for water policy, the fact that water planning regions do not have any particular social or political validity should be borne in mind. Often there is an urgent need to define water regions as something other than river basins. The concept of "problem-sheds," where small areas with common water problems are aggregated into larger regions, makes more sense. The Great Lakes Council, which coordinates research and water policies among the several states in the watershed of the Great Lakes, is a good example of this approach. Phosphate reduction in Lake Erie is an example of realizing economies through such a coordinated approach to common problems by several jurisdictions.

Intergovernmental relations. Relations between the federal water agencies and the states need clarification and policy reform. Part of the problem lies with the recent rise of state competencies and responsibilities at a time when federal programs have generally been on the decline. Although the "New Federalism" of the Reagan administration was in many ways a recognition of underlying facts of political life, it points to the larger policy question perennially at hand: what is the fundamental role for the federal government in water resources within a system of governance where the states have sovereignty over water? The clash of roles is especially apparent in the West, over the issue of reserved federal rights to water.

Institutional reform. The programmatic water activities of the federal government have one characteristic in common. They each require a central, facilitating institution of some sort. For the better part of a century, policy experts have been searching for just the right federal water resources institution. As early as 1908, for example, President Theodore Roosevelt's Inland Waterways Commission called for a new agency to coordinate the work of federal agencies in making multipurpose plans for the nation's waterways in cooperation with state and local governments—a recommendation that is still plausible today.

However, today there exist 34 federal agencies making water decisions, including 11 independent federal agencies in 10 cabinet departments, four agencies in the Executive Office of the President, five river basin commissions, the federal courts, and two bureaus. There are at least 25 separate water programs, governed by more than 200 sets of federal rules, regulations, and laws.

The federal executive branch institutions dealing with water possess a 90,000-person labor force, the bulk of whom are in agencies with shrinking budgets and outdated missions. Current and future-oriented missions need to be designed, and federal concern with water must be made proportional to federal concern in other areas of natural resources, such as the atmosphere, lands, and forests. This concern could be rectified by a reallocation of federal manpower to these areas.

The nation's primary water policy–making body, Congress, is equally fragmented. By the 102nd Congress, there were 14 House committees with 102 subcommittees, plus 13 Senate committees with 82 subcommittees, exercising responsibility over some aspect of water resources. Seventy-six separate congressional appropriations accounts for water have been identified. It is no surprise that the legislative enactments over the years have exhibited overlap, duplication, and even inconsistency.

In reforming institutions, Congress itself should not be overlooked. The problem here is manifest in three forms. The first is the proliferation of congressional committees that have something to do with water. The congressional practice of authorizing remedial efforts by specific agencies, in individual legislative acts, framed by different authorizing committees, and in largely prescribed, functional terms, makes coherent and efficient policy implementation by the executive difficult. The second is the proliferation of legislative staff. As a whole the U.S. Congress now employs 31,000 people, making it the most heavily staffed (some would say overstaffed) legislative body in the world. Thus, in water as in other fields a whole new federal bureaucracy has arisen on Capitol Hill, often with impressive professional credentials of its own, and so attuned to the political reward system as to be increasingly unwilling to relegate program definition responsibility to federal agency managers. The third is the tendency of such committees to micromanage agency programs through either highly prescriptive pieces of legislation or overzealous oversight activities. In many instances (e.g., the Safe Drinking Water Acts of 1974 and 1986), executive branch agencies are simply told what to do and how to do it. Having had little input into the prescriptions, they are nevertheless held accountable for implementation.

The obvious solution in each instance is a measure of consolidation: reform of the committee structure in Congress, and reorganization of the executive branch agencies. As with executive branch fragmentation, the most logical solution to the problem would be reorganization and consolidation actions by Congress itself, but most observers doubt that any real restructuring will take place in the foreseeable future. Barring that possibility, the next best approach would be efforts at improved coordination among joint authorizing committees of the House and the Senate, coordination that is generally practiced by state legislative bodies but not by the federal Congress. Sequential referrals of legislation to affected committees could improve coordination but at the risk of slowing down the legislative process. Select committees of either or both houses could be constituted to look at major policy or program issues. But there is one other parallel with the situation in the executive branch that is worthy of attention: the absence of a single point of responsibility within the legislative branch for reviewing and recommending federal water policy.

There has been talk about the need for coordination since the days of the New Deal. Various devices have been suggested or employed: voluntary interagency mechanisms, formal consolidation of agencies, the creation of specific coordinating committees, commissions, and the promulgation of uniform principles, standards, and guidelines governing water projects or procedures. All the institutional efforts toward coordination have encountered serious difficulties over the years.

Interest groups. Further complicating the process of achieving an orderly and comprehensive federal water policy are numerous outside interests that have strong, often vested concerns for the way federal water resources responsibilities are carried out. Among the most prominent are the 50 states, which in many instances are required to play a direct role in advancing the provisions of federal law. Economic interests in transportation, agriculture, and manufacturing are affected, both directly and indirectly, by federal water policies and projects, and take an active role in shaping them.

Special interest groups in recent years have moved away from besieging agency administrators in favor of direct dealings with members of Congress, their committees, and their staffs. Blocked in the executive branch by the nonactivist Reagan administration for eight years, the representatives of such groups found Congress to be more responsive. Ways will have to be found to make sure that the interest groups and their legislative and administrative allies allow for other less narrow views to intrude into their cozy world. Chapter 7 explores the relative impacts of some of the interest groups on the formulation of the major federal water legislation and its implementation.

Unrealistic expectations.   Overarching all aspects of the federal water policy scene are the unrealistic perceptions of the public. Americans expect to receive water of the highest quality, at the lowest price, and in unlimited quantity—everywhere and all the time. A New York Times/CBS poll of August 1990 revealed that fully 74 percent of the people surveyed agreed with the statement: "Protecting the environment is so important that requirements and standards cannot be too high, and continuing environmental improvements must be made regardless of cost." Only 21 percent disagreed. The "pro-environment at any cost" responses have consistently gotten stronger since 1981 (when only 45 percent agreed), when the question was first asked. These unrealistic expectations can only be dealt with by education. Although there are plenty of educational efforts in the water area, most of these miss the real targets. Looking toward the future, it is important to educate schoolchildren about water and its role in human ecology (which should include coverage of economics, finance, and issues of risk and uncertainty), but those who critically need the education are the politicians, political elites, and the general adult (that is, voting) population.

Public health versus environmental health.   In particular, the issue of how to deal with the riskiness of improving water quality needs to be debated. Is the goal improved human health? Is it sustainability of the ecosystem? How much can we afford? The public, the politicians, and the administrators of government policy must all be educated in how to frame, debate, and resolve these issues. The current debate in this area has been usurped by environmental extremists, who may have gained a less safe overall environment because of unrealistic demands on some part of it, like the draconian requirements of the Safe Drinking Water Act of 1986 (enacted in the absence of any documented epidemic threat to public health from drinking water).

Given complicated environmental legislation and the present U.S. culture of litigiousness, it is only natural to expect large numbers of lawsuits. The U.S. tort liability system has become an environmental control mechanism in its own right. Is this desirable? If so, what should be the federal government's role in providing environmental insurance or limiting liability?

In addition to the broader policy concerns discussed above, there are four specific substantive concerns, groundwater, non-point sources of pollution, wetlands, and drought, that figure in the unfinished business of federal water policy.

Groundwater protection.   With groundwater serving as the principal source of drinking water for 50 percent of all Americans, and for an estimated 97 percent of those living in rural areas, heightened federal involvement seems warranted for public health reasons if for no others. The independent National Groundwater Policy Forum, under the chairmanship of former Arizona governor Bruce Babbitt (now Secretary of the Interior), echoed the urgency in its 1987 report, *Groundwater: Saving the Unseen Resource.* It recommended a new federal statute establishing groundwater protection as a national goal, but with the states serving as the parties primarily responsible for implementation.

But despite a substantial consensus on the need to move ahead, the usual problems in federal water policy impede remedial action. Lacking a central water policy vehicle, the Reagan administration was unable to reach agreement within the federal bureaucracy on agency roles and responsibilities. Many experts assert that state and local governments, rather than the federal government, should exercise primary responsibility. Jurisdictional wrangling during the 100th Congress led to five separate committees of the House reporting their own versions of groundwater legislation. Unable to reach agreement on groundwater management, the measure finally considered was limited to data gathering and research. It died without passage, and no groundwater legislation was passed during the ensuing Bush administration.

Non-point source pollution.   The federal government will have to take a strong leadership role in helping the states implement coherent plans for non-point source pollution control. Distinctly different from the "end-of-the-pipe" discharge problems, non-point pollutants are typically dispersed substances like pesticides, fertilizers, and other agricultural chemicals associated with the use of rural or suburban lands; the sediments derived from mining, forestry, or construction activities; and, in urban areas, the wide range of pollutants associated with storm water runoff from city streets. Unlike point source pollution, which can be pinpointed, permitted by the EPA, and treated, non-point source pollution is best controlled through adjustments to the activity itself. In many cases, the most effective non-point source controls are land use controls.

The non-point source pollution problem strikes at the heart of many of the policy dilemmas set forth above. The responsibilities affect a number of federal agencies. There is uncertainty over the respective roles of the federal, state, and local actors. There is substantial overlap in the legislation (e.g., the related provisions of the Food Security Act of 1985, whose Title XII mandated the preparation of formal conservation plans for the nation's 1.5 million farm owners by 1990, and calls for plan implementation by 1995). In short, the situation seems tailor-made for White House leadership committed to a more cohesive federal water policy.

Wetlands. Since the advent of the first European settlers, more than one-half of the wetlands of the United States have been drained and converted to other uses.[2] Over the past 20 years wetlands conservation and regulation have become increasingly important issues. The National Wetlands Policy Forum's goal of "no net loss" (1988) was espoused by President Bush and his EPA administrator to little avail. During its tenure, the Bush administration was continually haunted by this sound bite. For over 100 years it was the explicit policy of the federal government to help drain the wetlands and swamps, and such traditions die hard. To achieve the new goal will require extensive revisions in the way federal, state, and local agencies deal with wetlands. Beyond the goal of "no net loss" there is now a widespread perception that the total area of wetlands needs to be increased, and the Forum recommended that the Corps of Engineers' mission be broadened to include the restoration and creation of wetlands.

Drought. Drought is relatively frequent in some parts of the country. It certainly is not so infrequent that prudent managers would not have contingency plans ready for when it occurs. The western governors were less than impressed by the federal government's preparedness in this area when they stated: "Indeed, each time the country experiences a drought, it has to create a mechanism for federal agency response, because no instrument of the federal government has been handed the assignment of thinking through in advance how federal agencies should address a drought when it occurs" (Western Governors Association, 1989, p. 4).

During 1988 a persistent drought gripped the central and southeastern sections of the United States. In the absence of a standing interagency institution, the president had to constitute a special task force of federal officials to monitor the drought and advise him of the remedial measures necessary. Less than 20 years earlier it had been the populous Northeast that felt the pressures of drought. In the West, water shortages have been a fact of life for as long as people can remember.

It is time to develop a national drought management policy and program comparable to the historic federal efforts in flood control. The program should deal with both pervasive and episodic drought. Caulfield (1988) pointed out that if the dire predictions associated with the greenhouse phenomenon came true, a large-scale federal effort could be required by the middle of the twenty-first century. Although the year 2050 seems a comfortable distance away, any attempt at long-range, drought-proofing solutions, such as converting lands to more drought-resistant crops like forest, would need to begin before this century comes to a close.

As indicated earlier, the recent California drought and the major 1988 drought contain seeds of opportunity for longer-term water policy reforms that are at the heart of this inquiry. For example, if improved interagency coordination is one such objective, the drought-related apparatus instituted in 1988 (and in 1991 in California) could be modified to serve other needs. Networking with state and local observers for drought reporting purposes might provide the essence of a larger water resource information and extension system. The need to assemble data related to drought conditions cannot help underscoring other needs for timely and accurate information and resource assessments. In response to the events of 1988, Congress requested the Corps of Engineers—under its mandate by the Water Resources Development Act of 1986—to coordinate a four-year national drought study. The study's final report, entitled "The National Study of Water Management during Drought," is due in October 1993 (Werick, 1993; U.S. Army Corps of Engineers, 1991).

Sustainability.   In recent years there has been much discussion of sustainability of human societies and ecosystems. Although the definitions are elusive, the concept has struck a chord in most perceptive observers. Is what we are doing sustainable into the future? If not, what can we, or should we, do to make it so? With respect to water use in the United States the major sustainability issue is how long we can continue to ignore in-stream flows of rivers. Reduction of flows by stream diversions, or change in the physical and chemical composition of the water due to return flows from industry, municipalities, and agriculture, can have serious consequences for the ecosystems that have developed under particular hydrological regimes. For example, the impact of the drought in California on fish and wildlife led to major changes in federal allocations of water from the Central Valley Project away from irrigated agriculture and toward ecosystem preservation. This is a major victory for the sustainability of the California aquatic ecosystems, but much more remains to be done in other parts of the country.

Outline of the Book

The book is essentially in three parts. The first part deals with the problem of water management from scientific, technical, and historical points of view. This first chapter has given an overview of the plethora of water issues we have before us. Chapter 2 covers the basic scientific and technical issues involved in understanding (1) the resource base, (2) how much of the resource is needed for various human and nonhuman activities, and (3) what underlies the standards set in regulations controlling water use. Chapter 3 gives a quick overview of the history of water policy in the United States, emphasizing the similarities of the problems over the centuries and our recurrent struggles with them—with and without resolution. It is generally not true that ignorance of history will force one to relive it, but it is certain that one can learn a lot from a careful study of history, particularly in the water policy arena. The first part ends with chapter 4, which discusses what makes water different from other natural resources.

The second part deals with active intervention in water policy. Chapter 5 gives the technologist's approach to resolving water issues: no problem, everything can be easily solved by more technical control of nature. Chapter 6 follows with the economist's perspective, which, like the technologist's view, sees little problem in resolving water issues provided one has mastered the correct pricing or resource allocation paradigm. Only in chapter 7 do we begin to get a sense of the real difficulty in making group decisions in a democracy about such a pervasive natural resource as water. This chapter makes it clear that we will always have water conflicts in the United States. There is no happy place somewhere over the rainbow where all will be resolved.

The third part of the book deals, in chapter 8, with the need for institutions that will reflect the technical, scientific, economic, and political aspects of water and that will integrate them in a way that will lead to socially acceptable and scientifically rational outcomes. Finally chapter 9 offers a blueprint for future water policymakers to follow.

Basic Hydrology, Water Demand, and Setting Standards

A Primer on Hydrology

The freshwater ecosystem. In an ecosystem everything is connected to everything else, which makes analysis particularly difficult. Freshwater ecosystems tend to be more difficult to analyze and understand than terrestrial ecosystems because of the complexity of the role of water in sustaining the flora and fauna (biota) of the system and in the transport and diffusion of chemicals and nutrients throughout the system.

The concept of the *hydrological cycle* is basic to understanding the freshwater ecosystem. Under this concept, the ecosystem is a natural machine, a constantly running distillation and pumping system. The sun applies heat energy, which together with the force of gravity keeps water moving. Although the cycle has neither beginning nor end, from our point of view the oceans are the major source, the atmosphere is the delivery vehicle, and the land is the user. In this system there is no water lost or gained, apart from a very small amount that is involved in chemical reactions and is removed from the water cycle as either hydrogen or oxygen. But the amount of water available to the user may fluctuate widely and critically because of variations at the source, or more usually in the delivery agent. Large temperature changes in the atmosphere and the oceans have produced deserts and ice ages; small local alterations of the patterns of the hydrological cycle produce floods and droughts.

The hydrological cycle (figure 2.1) accounts for the passage of water in its various phases through the lithosphere (from the surface down to the mantle rocks) and troposphere (from the ground up to 12 kilometers into the atmosphere). Aquatic ecosystems exist wherever water concentrates in this cycle—in lakes, rivers, marshes, oceans,

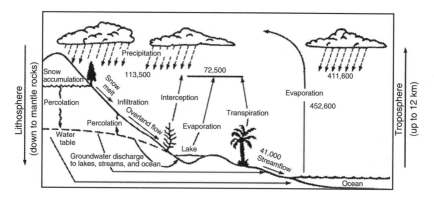

Figure 2.1
The global hydrological cycle, showing approximate magnitudes of components in km^3. (Source: Repetto, 1985. © 1985 by Yale University Press.)

in soil moisture, and in groundwater. The three most important stages of the cycle are evaporation, condensation, and precipitation. The atmosphere acquires moisture by evaporation from oceans, lakes, rivers, and damp soil, and by transpiration from plants. These processes are referred to jointly as "evapotranspiration." Air currents transport the water vapor over large distances. Condensation, cloud formation, and precipitation occur. During precipitation, there is some reevaporation in the air but much of the water reaches the ground, a water surface, or vegetation. Of the precipitation that reaches the vegetation, some is held by the canopy of the trees and eventually evaporates again, some will run down, drip from the canopy, or be shaken off by the wind and reach the ground. Of the water that reaches the ground, some will accumulate in surface depressions, some will begin to move over the surface, some will infiltrate below the surface and percolate to deeper layers into the groundwater, where it may be held for long periods. For example, water in some of the aquifers under the Sahara Desert is millions of years old; this water is essentially a nonrenewable resource since there is effectively no recharge to the aquifer under the present climate.

Climate shapes the physical and biological environment, which in turn influences the composition and state of the atmosphere; we see the strong interactions and feedback loops that are characteristic of ecosystems. Human activities are influenced greatly by weather and climatic conditions and subsequently exert an equally great influence (either purposefully or inadvertently) on weather, climate, the moisture cycle, and the geochemical cycles. Many climatologists believe that twentieth-century developments are the source of changes in the carbon cycle, specifically an accumulating increase of CO_2 in the atmosphere and the hypothesized "greenhouse" effect whereby the global temperature could rise by as much as 3°C by the middle of the next century.

Variability in precipitation has great significance for the stability of the ecosystem. In semiarid areas it can be devastating. Average annual rainfall worldwide ranges from almost zero in some deserts to over 10,000 millimeters (almost 40 feet) in parts of Hawaii. At Iquique in the Chilean desert, no rain fell for a period of 14 years; at the other extreme, 26,466 millimeters (86.83 feet) were recorded in a single year (August 1860–July 1861) at Cherrapunji in India. In the Sahel region of Africa during the late 1960s, a decrease in rainfall of 20 to 50 percent over a six-year period seriously affected agriculture, livestock, inland fisheries, and entire national economies, with many social and political ramifications. The causes of such droughts are not known, nor is it understood whether a decrease in rainfall represents an irregular fluctuation within an unchanged larger pattern or is part of a permanent long-term change in climate.

The global water resource base.   The most comprehensive studies of the global water balance are those by Russian geographers, based upon the original work of L'vovich (1979). Even though L'vovich's studies are based upon data from the 1950s they are still considered the basis for most useful global estimates. His major results are shown graphically in the hydrological balance in figure 2.1. L'vovich estimated the total annual precipitation on the entire globe to be 525,100 cubic kilometers (1 km³ is equivalent to 264.2 billion U.S. gallons; the annual average flow of the Colorado River at Yuma, Arizona, is about 12 km³).

Seventy-eight percent of precipitation is over the oceans and therefore not readily available for human use. Only 7.8 percent of the precipitation is available annually as surface or groundwater runoff, and only 2.6 percent is available as stable runoff;[1] the rest, varying greatly from year to year, flows to the sea as flood runoff. Of the total precipitation, only 1.7 percent or 9,000 km³ is readily available for human use. Worldwide diversion of water for human use in 1970 was about one-third of this amount, and it has been estimated that as much as two-thirds of the total readily available worldwide is affected in some way by anthropogenic pollution (L'vovich, 1979). Hence the percentage of available resources actually used by man, including allowances for dilution of wastes, appears to be in the range of 38–64 percent of the readily available resource, though it is only 1 percent of the total precipitation.

These figures indicate that currently the global water resource does not appear to be scarce. If, however, a doubling of use occurred over the next 20 years and nothing was done to increase the stable runoff and reduce water pollution, then the worldwide resource could become severely stressed. Moreover, despite the apparently adequate aggregate resource, there are many serious scarcities even at the current level of human use at regional and individual country levels (see Falkenmark, 1989).

What can be concluded from these data on global water resources? Despite enormous quantities of water available in the total hydrosphere, the actual amounts available on a sustained basis for human use in the populated regions of the globe are surprisingly close to the potential "needs" based upon historical usage. Unless we start to moderate these needs by proper planning and management, as is now beginning to happen in the United States, Japan, and Europe, the situation could become quite serious during the next 50 years.

The U.S. water resource base. Assessing the resource base of a country as large as the United States raises some serious conceptual problems. The first problem is how to find out how much water is entering the geographic boundaries of the country. By using rain gauges it is possible to get a good idea of how much rain falls on a particular spot, but putting in enough gauging stations to measure and to collect data over a large country is very expensive. Many would argue that the number of rain gauges currently in the federal networks (more than 6,000) is not adequate.

The flow of water in U.S. streams is also recorded via a system of 6,800 continuously recording sites around the nation (U.S. Geological Survey, 1986, p. 98). In addition to measuring daily flow rates, the quality of the water in the streams is also measured on a continuously recording basis through a separate network of 550 stations around the country. The EPA's STORET system reports on between 500,000 and 600,000 sites for surface water quality and about 150,000 for groundwater quality at varying frequencies and measurement inten-

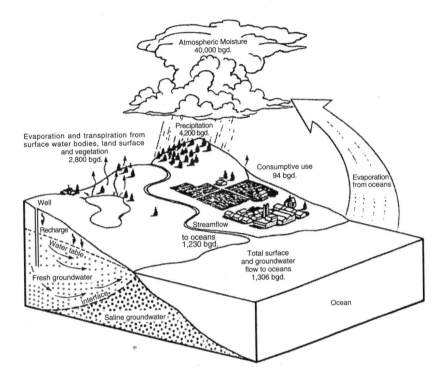

Figure 2.2
Estimated use of water in the United States, 1990. The hydrological cycle show-
ing the gross water budget of the coterminous United States. (Data from Solley,
Pierce, and Perlman, 1993.)

sity. Which water quality parameters to measure is a major unresolved question. Those that are now being measured are mainly those
that are easiest to measure, particularly in the fully automatic stations
where only limited chemical and biological data can be easily recorded and transmitted.

Figure 2.2 shows the best available estimates of the annual gross
freshwater budget of the coterminous United States (Solley, Pierce,
and Perlman, 1993). Out of an atmospheric moisture passing over the
country each year—averaging 40,000 billion gallons per day (bgd)—
about 10 percent falls as precipitation. Of this, 67 percent evaporates
and transpires from surface water bodies, the land surface, the upper
reaches of shallow groundwater aquifers, and vegetation. Of the remaining supply, 29 percent returns to the oceans as streamflow, 2
percent returns to the oceans as drainage out of the groundwater aqui-

fers, and only 2 percent goes to consumptive use by man (slightly over 94 bgd).

On the basis of the above one could assume that the United States could safely use, on an annual basis, up to the amount that does not evaporate, i.e., 1,400 bgd. Since we only consume 94 bgd the conclusion would seem to be that there is no possibility of water shortages. However, the average figure masks four important facts that make meaningful assessment of the resource extremely difficult and invalidate the above conclusion.

First, human use is reported as consumptive use; this is the correct measure to use when dealing with a renewable resource, but the actual diversions of water run about 400 bgd. Most of this is returned to the watercourses after only a short time, but it is often degraded in quality and returned at a different location; e.g., extracted groundwater used by municipalities is typically disposed of into the nearest surface stream, from where it may or may not return to the groundwater. This means that even though it is theoretically returned to the cycle for use it may not be of practical or economic use during that year and in that part of the country.

Second, the interyear stochastic fluctuations in precipitation diminish the average figure of 1,400 bgd for the average year to 675 bgd in a 5th percentile year (a year whose precipitation is exceeded in 95 out of 100 years, a typical reliability level for water supply).

Third, the availability of rainfall and surface water varies very widely by geographic region. While on average there may be plenty of water for all conceivable uses well into the future, there may be large parts of the country facing serious shortages. Adding the seasonal variability and the random nature of the precipitation to this picture, severe water crises become possible even in this well-watered country.

Finally, the fourth and most often neglected point is one of great ecological importance. Water that runs to the sea is not "wasted." Much as we might like to, we cannot simply dry up the rivers; instream water requirements are substantial and have to be met. The breeding habitats of fish and other wildlife will be destroyed if adequate water flows are not maintained in the rivers and estuaries. Only recently have federal and state laws begun to recognize such in-stream uses as "beneficial uses."

The existence, however, of large developable groundwater aquifers in the United States means that by judicious use of the groundwater resources one can actually plan closer to the mean resource availability rather than for the lower extremes. This is because groundwater acts like a flywheel in the system; when the precipitation is low the water can be taken out of the ground, and when the precipitation returns to its former levels it will usually replenish the aquifer. The crucial question here is the rate of replenishment of the aquifer. For example, in parts of the Ogallala aquifer, which runs from Texas to Nebraska, the rates of recharge may be as low as 5 mm per year, which implies that it would take hundreds of years to recharge the badly overdrafted regions. Nevertheless, in other parts of the United States the recharge is orders of magnitude faster and the groundwater does serve to balance out interannual fluctuation in available precipitation.

The reliance on groundwater for supply leads directly to problems with groundwater contamination. While it is no more difficult to clean up groundwater than surface water for human or other uses, groundwater contamination raises many more fears in the general population than similar levels of contamination in the surface water. Some of these fears stemming from the unseen nature of the contamination are well founded because many persons still rely on self-supply from untreated well water.

Uses and Quantities of Water Demanded

Because water has been treated in noneconomic ways in the past it is difficult to assess the demand for water in its technical economic sense, that is, the quantity that will be purchased at a given price level. All too often water is viewed as a "need" and amounts currently "needed" are projected into the future. If water has been widely wasted in the past, then projecting the same behavior into the future leads to the expectation of huge requirements and seeming conflicts with the resource base. The traditional way to estimate "demand" is to take current amounts of water supplied for each type of water use and increase these figures in proportion to the projected levels of population and economic growth. These projections, however, ignore the strict economic interpretation of consumer demand as being based upon the price of the resource and the level of income of the population, among other factors.

Quantities used in the United States. The water dilemmas in the United States are represented by figure 2.3. Irrigated agriculture, located mostly in the 19 western states (the reclamation states), con-

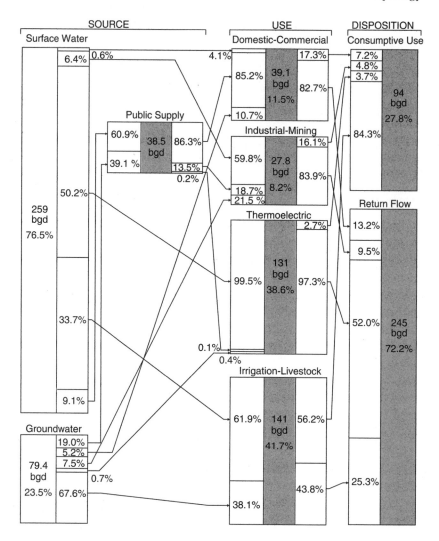

Figure 2.3
Estimated water use in the United States, 1990. Freshwater withdrawals and
disposition of water in billion gallons per day (bgd). For each water use category,
this diagram shows the relative proportion of water source and disposition and
the general distribution of water from source to disposition. The lines and arrows
indicate the distribution of water from source to disposition for each category;
for example, surface water was 76.5 percent of total fresh water withdrawn, and,
going from "Source" to "Use" columns, the line from the surface water block
to the domestic and commercial block indicates that 0.6 percent of all surface
water withdrawn was the source for 4.1 percent of total water (self-supplied
withdrawals, public supply deliveries) for domestic and commercial purposes.
(Source: Solley, Pierce, and Perlman, 1993.)

sumes four times as much water as all other users of the nation put together. The asymmetry of this water use by small numbers of farmers in the West, often with large federal subsidies, is a cause of serious concern over the equity of federal resource allocation in water resources. It has been divisive in Congress, tending to raise issues of regional chauvinism rather than rational debate. In 1982, commenting on the Reclamation Reform Act which still maintained large federal subsidies for irrigation water, Senator Daniel Patrick Moynihan of New York said that the bill represented "all that is wrong with federal involvement in water resources: chaos, arbitrariness, inequity, and waste." The senator, however, changed his tune in 1986 when he led the charge to overturn President Reagan's veto of the Clean Water Act, which incidentally included hundreds of millions in federal subsidies for wastewater treatment plants in New York state.

Tables 2.1–2.3 show estimates of water use by residences and selected industries. It is interesting to note that of the water use in a typical family residence the bulk goes to toilet flushing and showering. Another way of looking at the numbers is to compute the amount of water required to produce $1,000 worth of product. Welsh (1985) shows that grains use 13.65 million gallons of water to produce $1,000 of output, citrus 3.9 million gallons, and manufacturing on average only 0.65 million gallons. There is no doubt as to which are the most economically productive uses of scarce water supplies.

Figure 2.3, based upon the most recent data (Solley, Pierce, and Perlman, 1993), illuminates in more detail the magnitudes and the origins of many of the water pollution problems. For example, the largest source of returned water is from the thermoelectric power industry. By and large, this returned water is not a serious contamination problem: it typically returns to the nearest water body four or five degrees Centigrade above the background temperature. Depending upon the ambient temperatures and the local aquatic fauna and flora, this generally is tolerable or can easily be made tolerable. Agricultural runoff, on the other hand, which is the next largest source of return flows, is the largest contributor of two serious water pollution problems: groundwater contamination and non-point source pollution. Recent attempts to pass federal groundwater legislation have been unsuccessful, but the 1987 amendments to the Clean Water Act took the first rudimentary steps to address the non-point source issue.

Industry, mining, domestic, and commercial users are the source of only 22.7 percent of the returned water, but together account for

Table 2.1
Typical Rates of Water Use for Various Industries

Industry	Range of Flow, (gal/ton product)
Cannery	
Green beans	12,000–17,000
Peaches and pears	3,600–4,800
Other fruits and vegetables	960–8,400
Chemical	
Ammonia	24,000–72,000
Carbon dioxide	14,400–21,600
Lactose	144,000–192,000
Sulfur	1,920–2,400
Food and beverage	
Beer	2,400–3,840
Bread	480–960
Meat packing	3,600–4,800[1]
Milk products	2,400–4,800
Whisky	14,400–19,200
Pulp and paper	
Pulp	60,000–190,000
Paper	29,000–38,000
Textile	
Bleaching	48,000–72,000[2]
Dyeing	7,200–14,400[2]

Source: Metcalf and Eddy, Inc. (1991). Used by permission of McGraw-Hill, Inc.; © 1991.
[1] Live weight.
[2] Cotton.

the bulk of conventional point source pollution. They have been the target of most of the water quality legislation and regulation, even though point source pollution is not the major pollution problem. The current emphasis, however, is on control of point source effluent quality rather than water quality.[2]

Finally, it is important to remember that the demands for water are not constant, nor are they always increasing as most people—even many water professionals—imagine. The recent history of water demand in the United States (figure 2.4) shows that the per capita consumption of water in 1990 was less than in 1965, despite an almost 50

Table 2.2
Typical Municipal Water Use in the United States

Use	Flow (gal per capita per day)		
	Range	Average	Percent Based on Average Flow
Domestic	40–130	60	36.4
Industrial (nondomestic)	10–100	70	42.4
Public service	5–20	10	6.0
Unaccounted system losses and leakage	10–40	25	15.2
	65–290	165	100.0

Source: Metcalf and Eddy, Inc. (1991), based on Tchobanoglous and Schroeder (1985). Used by permission of McGraw-Hill, Inc.; © 1991.

Table 2.3
Typical Distribution of Residential Interior Water Use[1]

Use	Percent of Total
Baths	8.9
Dishwashers	3.1
Faucets	11.7
Showers	21.2
Toilets	28.4
Toilet leakage	5.5
Washing machines	21.2
	100.0

Source: Metcalf and Eddy, Inc. (1991), based on U.S. Department of Housing and Urban Development (1984). Used by permission of McGraw-Hill, Inc.; © 1991.
[1] Without water-conserving fixtures.

percent increase in real per capita income over that period. The *total* demand for water also continued to decrease from the 1980 high, although 1990 was slightly higher than 1985; 1990 withdrawals were down almost 7 percent from 1980, and the absolute level was almost identical to that of 1975.

These changes in water use patterns need to be carefully studied in terms of who was doing the saving, and in which parts of the country. Large reductions in water use by agriculture and industry appear to have occurred between 1980 and 1990. From a policy point of view this change is of fundamental importance. To get an idea of its magnitude, compare the Water Resources Council's 1978 forecast

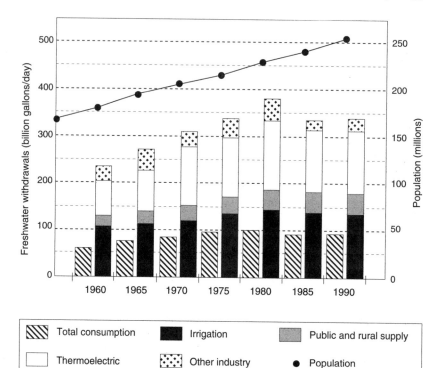

Figure 2.4
Trends in U.S. freshwater withdrawals and consumption, 1960–1990. (Data from
Solley, Pierce, and Perlman, 1993.)

of consumption of fresh water for 1985 (120 bgd) with the 1988 USGS's
estimate of actual 1985 consumption (94 bgd). This implies an overestimate of 30 percent—an error with immense planning implications, especially concerning the provision of water and wastewater
services. This gap between forecast and fact is likely to increase to an
overestimate of 43 bgd by the year 2000, almost half of the amount
currently supplied.

Uncertainty in Resource Base and Demands

Data about the water resource base and the demands placed upon it
are often given as though these numbers were exact and empirically
measured. In fact, resource base data are based upon extrapolation
from a limited number of observation points for a limited number of
years of daily, weekly, or even monthly measurements. Hydrological

and climatological data are notoriously variable, with the greatest variability being in arid regions with the smallest available resource. For example, the amount of rainfall available in 95 out of 100 years in the desert in Arizona is much less than one-half the average rainfall, whereas the same 95 percent reliable rainfall in Massachusetts is more than three-quarters of the average rainfall.

In addition to these kinds of uncertainty there are other measurement problems. For example, whereas streamflow data are measured at over 6,500 stations on the 1.5 million miles of the nation's rivers and rainfall is measured at over 6,000 locations over the nation's 3.5 million square miles, evaporation (which accounts for the largest component of the water balance) is measured at fewer than 100 stations, and there is *no* systematic national groundwater monitoring program—the resource that supplies 50 percent of domestic consumers. Moreover, even where observations are made they tend to be very selective of the actual quantity and quality parameters measured.

One anomaly in defining the amount of water available is the fact that the actual development of the water resource may expand the amount of resource available by the simple mechanism of speeding up the hydrological cycle and rearranging the relative magnitudes of the component flows. Groundwater development in the Central Valley of California and the use of the aquifers of Florida over the past century are two cases where this has happened. The Central Valley case is a dramatic example of the dynamic effects of human interference with the hydrological cycle. Originally the Central Valley aquifer system was in a low-level steady-state equilibrium. Recharge to the aquifer was 1.8 bgd and the natural discharge was 1.8 bgd. The current situation, with the same net amount of water coming in as precipitation, now achieves a recharge of 9.8 bgd (a fivefold increase) and currently supports the pumping of 10.2 bgd. The natural discharge to the ocean has dropped to 0.3 bgd and there is a decrease in the groundwater stored in the aquifer equivalent to 0.7 bgd. The Florida case indicates that the introduction of 3 bgd of pumping only decreased the natural discharge by 1 bgd. The moral here is that it can be quite easy to be misled by the current definition of resources; careful water balances need to be estimated for the idiosyncrasies of each case.

Another example of this is the definition of drought. Some authors define a "meteorological" drought (when the precipitation is less than usual), others use an "agricultural" definition (when there is not enough precipitation for crops), and others a "hydrological" definition (when less water is available than usual from the sum of precipitation and streamflow). The difficulty arises from the fact that applying these three definitions can give radically different results depending on the nature of human interaction with the hydrological cycle. The provision of surface storages, the development of groundwater, and the changing of the agricultural cropping patterns all have a major impact on whether a dry period is a drought or not.

Physical and biological data are expensive to measure and collect, but the social science data dealing with a consumer's water use behavior are even more difficult to conceptualize, measure, and collect. In many cases only rudimentary notions of the magnitudes of these data are available to decision makers attempting to formulate and implement water policy.[3]

Setting Standards for Water Quality

Over most of the period since the first colonial settlements in America, water's role in economic development was a much lesser concern than its role in protecting human health. The great irony in the historical development of water policy is that the federal government only took a serious interest in water quality issues when, by and large, the serious health problems associated with water pollution were already solved by the municipal and local authorities. The federal government paid virtually no attention to the quality side of the water environment until the founding of the Public Health Service in 1912. There was no federal intervention despite overwhelming epidemiological evidence of the dangers to human survival. When the federal government did become involved in 1912, its initial concern was with the quality of drinking water served in interstate commerce (the commerce clause of the Constitution), and later, in the 1940s, with ambient water quality in lakes and rivers.

Dealing with water quality has turned out to be significantly more difficult and more expensive for the federal government than developing and regulating the quantity of water supply. To deal intelligently with quality, it is necessary to have a detailed understanding of the underlying chemical and biological processes involved, yet we do not have a complete understanding of the fate of chemical pollutants or their impacts upon humans and animal species in the aquatic ecosystem. Consequently, some major scientific and technical assumptions must be made in setting standards for water quality.

A fundamental question is whether standards should be geared toward human use of the water, human health impacts, or the quality of the ambient environment. Does the ecosystem have any rights other than sustaining itself for continued human use? Most people eventually end up as *Homo sapiens* chauvinists despite the pro-ecosystem preamble of the Clean Water Act of 1972 and its amendments. If human use is the issue, which uses should be permitted? If health is the issue, how safe is safe? Water quality standards for toxic chemicals are typically based upon laboratory tests of the effects of contaminants at high concentrations, over a short period of time, on laboratory animals. There is little actual epidemiological data on long-term low-level exposure for humans. The current emphasis on very stringent standards for drinking water comes at a time when there is *no epidemiological* data supporting the idea of large or significant public health risks due to drinking water.[4]

The lack of epidemiological data is not sufficient to say that we should not care about the quality of water; if there were other evidence to show that humans were being exposed to dangerous conditions, it might be prudent to reduce this risk. Many environmental groups talk about the "drinking water crisis" as though it were a generally agreed upon fact, but the opposite seems to be true—the quality of drinking water available to the American public is better now than it has ever been. There is no public health crisis due to water consumption. Why then the crisis atmosphere? The facts are on face value intimidating:

• Over the past 50 years mankind has introduced many thousands of newly synthesized chemicals into the environment, many of which are carcinogenic or otherwise harmful to mankind.
• These chemicals can now be measured in trace quantities in the air we breathe, the water we drink, and the food we eat (not surprisingly, the better our ability to measure trace amounts in parts per billion, the more often we find the chemicals).
• Groundwater (which accounts for 50 percent of domestic consumption) is quite often contaminated with these chemicals and other naturally occurring ones.
• There have been several widely reported cases of egregious disposal of hazardous chemicals in the environment (most notoriously at Love Canal in New York state).

Coupled with serious scientific concern about the fate of pollutants in the environment in the mid-1970s, the above facts helped create a crisis atmosphere with respect to human health and water contamination.

If ambient quality of the environment is at stake there is little or no guide for reasonable risk levels. The "seven-day low flow exceeded once in ten years" has traditionally been used as a basis for calculating worst-case waste dilution scenarios for setting ambient water quality standards—not drinking water standards, but the quality of water in streams, lakes, and reservoirs. In many settings this implies a 99.99 percent level of reliability in meeting the standard (a one in 70,000 chance of violating the standard over a human lifetime). Is this too stringent, or not stringent enough? Obviously it depends to a large extent on the consequences of violating the standard. In human health terms it may mean death—what does it mean in terms of the environment and functioning ecosystems?

The ability to sense and avoid harmful environmental conditions is necessary for all living organisms. Most of us act instinctively to minimize our risks. In addition, we expect society to minimize the risks suffered by its members, subject to overriding moral, economic, or other constraints.

More than 20 years ago, Chauncey Starr, one of the early researchers in the area of weighing risks and benefits, assumed that by trial and error over time society had arrived at an "essentially optimum" balance between risks and benefits for all activities. He concluded that the acceptability of risk was proportional to the third power of the benefits, and that society would be willing to accept risks from voluntary activities, such as skiing, that are a thousand times as great as involuntary risks, such as contaminants in the drinking water, with approximately the same level of health benefits (Starr, 1969).

But even within the area of involuntary risk there appear to be inconsistencies not easily explained by Starr's theory. More recent commentators (Waterstone and Lord, 1989) contrast the ease with which society accepts the involuntary risks associated with dam and levee failures designed according to the "standard project flood" by experts with society's reluctance to accept a "one in a million" lifetime cancer risk level associated with drinking water quality standards.

By the end of the Reagan administration, the senior scientific staff of the EPA complained that "much of the agency's money and effort is devoted to managing lower-level risks like toxic waste dumps, underground storage tanks, and non-hazardous municipal waste sites" (Passell, 1989). But the results of several public polls showed that "EPA's priorities appear more closely aligned with public opinion than with our [own] estimated risks" (U.S. Environmental Protection Agency, 1987).

Comparing and contrasting risks.   Risks should be compared with each other for similar levels of health benefits. For example, chloroform, which is produced as a by-product of water chlorination, is often an accidental pollutant of water. It may be compared with trichloroethylene (TCE), an industrial solvent and also often an accidental pollutant of water. Based on animal tests, chloroform produces cancer 20 times as readily as TCE, although neither is known to cause cancer in humans; if they did, chloroform would be presumed to do so 20 times more readily (Crouch, Wilson, and Zeise, 1983). Given these risk levels one might reasonably expect that chloroform would have a much lower allowable level in drinking water. However, this is not the case, and indeed the reverse holds true. EPA's interim maximum contaminant level (MCL) for all trihalomethanes (mostly chloroform) is 100 ppb, while the recommended maximum contaminant level (RMCL) for TCE is 5 ppb.

   This sort of Alice-in-Wonderland regulation came about because if the chloroform standard had been set to conform with the TCE standard, very few water treatment plants in the nation could meet the federal standard. This would clearly not be a politically acceptable outcome. Is the public health compromised by such regulatory behavior? Some environmentalists argue that it is, but toxicologists argue that comparison with similar contaminants is not the only way of looking at the issue, and that rather the risk should be *contrasted* to other involuntary risks that are controlled or controllable by the federal government.

   In contrasting risks, one needs to look at other involuntary risks in other areas. For example, contaminants in water should be juxtaposed with risks in food additives, food processing, occupational exposure, and naturally occurring toxic and carcinogenic materials. When this is done for cancer risk (Ames, Magaw, and Gold, 1987), the risks associated with both TCE and chloroform appear minor in comparison to breathing the air in a conventional home, which is calculated as being 600 times more hazardous than either one of the above because of indoor air pollution. Similarly, aflatoxin and dioxin, two of the most carcinogenic chemicals known, have similar toxicities and carcinogenic potency, yet standards are much more stringent for dioxin than for aflatoxin. The explanation may be as simple as the fact that dioxin is a man-made chemical, and we are much more suspicious of man-made toxins than those that occur naturally. Clearly, how stringent we should be depends upon who is being asked, who is at risk, and who bears the cost.

Costs of meeting regulations. There are no hard figures on exactly how much it costs to meet environmental standards, but Abelson (1993) reports some estimates. He estimates that the annual cost amounted to $115 billion in 1991 and was likely to double by the year 2000. Abelson reports that city leaders are complaining most about the cost of meeting water regulations. For example, the EPA has estimated that the nationwide capital cost of its proposed radon in drinking water standard, of 300 picocuries per liter, would be $1.6 billion. But the Association of California Water Agencies found that the costs would be $3.7 billion for California alone and between $12 and $20 billion nationwide, and these costs would only reduce public exposure to radon by 1 percent. There is increasing concern that the regulation of environment has lost touch with the real costs and benefits of environmental quality.

Improving risk perception. The most frequently advocated recommmendation for broadening the public understanding of risk is to present quantitative risk estimates for a variety of hazards in some unidimensional index of death or disability. Research suggests, however, that these sorts of comparisons may not be very satisfactory. "Riskiness" means more to people than "expected number of fatalities" or some other index of hazard. "Some of these debates may not even be about risk. Risk concerns may provide a rationale for actions taken on other grounds or they may be a surrogate for other social or ideological concerns. When this is the case, communication about risk is simply irrelevant to the discussion. Hidden agendas need to be brought to the surface for discussion" (Slovic, 1987, p. 285).

Because of the large number of low-level carcinogenic and other hazards and the high costs of regulation, it is important not to divert society's attention away from the few really serious hazards (such as tobacco or saturated fat) by the pursuit of hundreds of minor or nonexistent hazards (Ames, Magaw, and Gold, 1987). And since human consumption of water is only between one and two liters per day, the animal evidence provides no good reason to expect that chlorination of water or current levels of man-made pollution of water pose a significant carcinogenic hazard. In the meantime the EPA is busy preparing regulations to control 83 contaminants in drinking water at the direction of the Congress.

This chapter has made an excursion into some of the important issues of water availability, water demand, and the setting of water quality standards. Taking all the uncertainties into account raises very severe problems for water policymakers. In many cases they are left having to balance risks with political pressures, and water planners and managers typically have little time for explaining the ambiguities involved in particular actions.

3
History of Water Policy in the United States

Ignorance of history does not mean that we are condemned to relive it, but it does mean that we deprive ourselves of the useful lessons that previous generations learned by painful trial and error. Almost all current proposals for changes in water policy are ideas that were either in use or proposed earlier. History teaches that despite a lack of scientific information, the old-time water policymakers generally made wise decisions given the situations they were in. Calls for the privatization of municipal water supply and wastewater management are really calls to return to the earliest periods in U.S. history when all such services were provided by the private sector. New ideas about setting up trust funds for specific types of water problems echo ideas current in 1902 when the Reclamation Fund was established to develop irrigation in the West. Many of the ideas of the current environmental movement are similar to those of the conservation movement of 80 years ago. A lot can be learned from looking back to see what worked, what was discarded, and why.

The history of U.S. water resources has been documented by various authors with different perspectives. Much attention has been given to the history of the 1930s and 1940s, and to specific agencies such as the Army Corps of Engineers (Maass, 1951) or the Bureau of Reclamation (Reisner, 1986). Comprehensive studies have been carried out by professional historians within the agencies themselves (Holmes, 1979, of the Department of Agriculture; and Reuss and Walker, 1983, of the Corps of Engineers). There is also a genre of recollections by major participants in the policymaking process.[1] One study stands out as the single most comprehensive study of water policy: the two-volume work by Holmes (Holmes, 1972, 1979). The first volume covering 1800 through 1960 is relatively short; the second covering 1961 through 1970 is more than three times as long, reflecting not only the heavy investments but also the enormous changes in policies and perceptions of water during that period. Appendix 1 gives a chronology of past federal actions and policies in the field of water, which also indicates the relative importance given to water policy in the different epochs. The presentation below follows closely the work of Foster and Rogers (1988), heavily influenced by Holmes (1972, 1979).

Early History of Federal Involvement

Even before there was a United States of America, water-related issues were prominent. In colonial Massachusetts, for example, the earliest enactment of the Great and General Court, as the legislature was known, was a measure to regulate the fishery—a matter, apparently, of even greater priority than the next legislative act, which was the establishment of courts of justice in the new world. In these early days, natural bodies of water 10 acres or more in size were declared to be "Great Ponds," open to all for fishing and fowling. Similar action was taken with respect to the so-called "foreshore"—the portion of the ocean shoreline lying between the high and low water marks. Not surprisingly, some of the earliest governmental functions, exercised by the colonies and later by the states, related to the regulation of harbors, waterways, and fisheries. Although the legislated matters were all very practical, since livelihood and survival were at stake, there was already a glimmering of the motivating forces that would dominate water policy in the future.

Among the earliest federal responsibilities was adjudicating differences among the states, of which many were water-related. Boundaries were one area of contention, particularly where borders ran offshore or along the thread of a stream. The progression from border disputes to resource disputes was only natural. As early as 1785, for example, we find Virginia and Maryland agreeing by interstate compact to enact concurrent fisheries regulations for Chesapeake Bay and the Potomac estuary. However, such compacts required the consent of Congress, and to this day the so-called consent provision of the Constitution ensures a federal say in virtually every interstate arrangement. In fact, it was the ratification of an illegal compact between Virginia and Maryland in 1785, the Mount Vernon Compact—arranged so that George Washington's Potowmack Company could build the Potomac Canal—that led to the Constitutional Convention in Philadelphia (Garrett, 1987).

One of the earliest manifestations of the tangible federal presence in water resources stemmed from Washington's favorable experience with engineers during the Revolutionary War, which led to the establishment of an engineering school at West Point in 1802. As this was the only domestic source of trained engineers until well into the nineteenth century, it was only natural that the army engineers would be called upon for civil works functions in the early days of the country. The first such appropriation by Congress involved the modest sum of $3,500 for the provision of lighthouses, buoys, and other navigation improvements in 1797: the classic Adam Smith role of government. By way of contrast, the most recent omnibus authorizations, the Water Resources Development Act of 1992 and the Fiscal Year 1993 Water Appropriation Act, call for a federal expenditure of $5.8 billion, down from the $16.5 billion of the 1986 Water Resources Development Act.

Emergence of Federal Responsibilities

Viewed over time, there have been three main thrusts to federal water policy during the past 200 years: economic development, coordination, and regulation. While these were discrete concerns at certain points, in later years they came together in public policy attention.

In the early days, for example, when the nation was young, growing, and still largely unexplored, the national goal was to encourage settlement, preferably by those seeking family-size farms. There was an abundant land base available in the public domain as a result of agreements with foreign powers and treaties with native American tribes. Financial and technical resource shortages at state and local levels decreed a growing role for the federal government in facilitating settlement and development.

Among the earliest development projects were those related to navigation. The objectives were to provide improved means of transportation and to open up avenues for commerce. The concept of a heightened federal role was not without controversy, however, for it flew in the face of the traditional, state-based view of federalism. An example of this was President Madison's rejection of the federal funding of the Erie Canal in 1811. Among the public policy benchmarks of this period were Treasury Secretary Albert Gallatin's report (1808) recommending a nationwide system of canals and other internal improvements (which served as the basis for the 1874 report of Senator William Windom's Select Committee on Transportation Routes to the Seaboard), and Chief Justice John Marshall's majority opinion of the Supreme Court that the federal power to regulate interstate commerce carries with it similar federal authority over navigation (*Gibbons v. Ogden*, 1824).

This Supreme Court ruling led to cautious involvement of the federal government in a variety of surveys and small-scale investments in navigation improvements between 1824 and 1849. In 1849 an act was passed in Congress allowing Louisiana to sell swamplands and use the revenues to construct levees and drains. The following year these privileges were extended to all the other states. The ensuing massive destruction of wetlands has come back to haunt the twentieth-century inhabitants of these regions.

It was inevitable that the initial concern for waterways would lead to a parallel role for the federal government in flood control. The extensive Mississippi River floods of the 1870s dramatized the need for such involvement, and Congress was persuaded to establish a Mississippi River Commission in 1879 to develop a comprehensive flood control program. Similar problems in the Sacramento and San Joaquin drainages led to the creation of the California Debris Commission in 1893. As technical consultants to Congress, it was only natural that the Army Corps of Engineers would play the dominant role in these early flood control enterprises.

There were political realities at work, too. The Republican commitment to a greater use of federal powers and authorities held sway. As land settlement and development took shape, supportive coalitions began to form, notably among southern and western water interests in the "dry half" of the country. These served as powerful engines for distributive politics and persuasive voices for regional development through federal project support.

A federal government role in irrigation was a direct outgrowth of the new nation's settlement policies. Statutes like the Homestead Act (1862), the Desert Land Act (1877), and the Carey Irrigation Act (1894) offered land grants as tangible incentives for settlement. The policies were not without precedent, for two centuries earlier the eastern portions of the country had been colonized with the help of liberal land grants to commercial trading companies from the British Crown. In colonial times, new communities as well were created by similar grants extended to groups of settlers by local legislative acts. But the important difference between the eastern and western land grants was that many of the western holdings were simply not arable without the provision of water.

Major John Wesley Powell's pioneering survey report of the arid lands of the West (1879) heralded what was to become a new era for federal involvement in irrigation. Under the Reclamation Act (1902), the Secretary of the Interior was authorized to locate and develop irrigation works in, ultimately, 17 western states. For the first time, Congress provided for a substantial delegation of program authority to an executive branch agency, the Reclamation Service, to accomplish the provisions of the act.

From a financial policy point of view the 1902 Reclamation Act was very interesting. It created a Reclamation Fund from the sale of federal public lands initially in 16 western states. The investments in irrigation were to be self-financing based upon repayment by the beneficiaries to the fund. At no point was the fund understood as a give-away of federal resources. Indeed, the Reclamation Service was initially not dependent upon Congress to initiate, plan, authorize, and appropriate funds for projects. Only in 1914 did Congress prohibit expenditures from the fund without appropriations by Congress. It was only the successive decline in revenues from sale of land and congressional advances of funds that led eventually to massive federal subsidies for the reclamation projects.

Ever since the Desert Land Act of 1877 led to massive land speculation in the West and to little sustained irrigated agriculture, the federal government was widely viewed as the right level of government to salvage the water situation.

The Progressive Movement and Conservationism

The period from 1901 to 1920 can be considered the period of progressive conservationism, whose goal was to promote the wise use of natural resources for all people and for future generations. The progressive movement, starting with President Theodore Roosevelt and ending with Wilson, supported strong government active in public utility regulation, timber conservation, and "trust busting." Hence they found interventionist water policies congenial. As appendix 1 shows, despite a period of war, this was an era of a great deal of federal action and policy development.

During this period an important federal initiative emerged in hydroelectric power, because facilities for navigation, flood control, and irrigation often required the storage of water and this, in turn, created opportunities for the generation of power. More importantly, from a populist political point of view the competition of low-cost public power was viewed as a natural way to regulate private power monopolies. Formal power provisions were added to reclamation law in 1906 and were also included in numerous special authorizations to the Army Corps of Engineers.

After a decade of strident public debate, the Federal Power Commission was created in 1920, with responsibility for licensing nonfederal power developments on navigable waters in the public domain, managing the sale of surplus power generated from federal dams, and conducting investigations of water resources generally. The early commission was interagency in character, with members drawn from the departments of War, Interior, and Agriculture. Its basin-wide surveys, first undertaken jointly but later solely by the Corps of Engineers, became a classic series of reports, the forerunners of a concerted federal commitment to basin-wide water resources planning.

There was a backlash against the progressives during the 12 years of Republican control from 1921 to 1933. Public power was one of the touchstones, and one of the first acts of President Harding was to terminate all work on the Wilson Dam on the Tennessee River (Holmes, 1972). Herbert Hoover, as Secretary of Commerce and later as president, maintained this opposition to public power but adopted many of the progressives' views on conservation. Even so, as can be seen in appendix 1, the period was a relatively quiet one for federal water policy.

The New Deal

The opposition to public power throughout this period may help explain the subsequent enthusiasm by the succeeding administration of Franklin Roosevelt in 1933. In the New Deal era of the 1930s federal water resources development reached its zenith. Widespread economic depression triggered an unprecedented use of federal powers, many related to water resources, to stimulate construction industries and to provide jobs. The Tennessee Valley Authority (1933) was one such example—a unique, independent government corporation charged with multipurpose development and management on a regional rather than a functional basis. The goal of attention to rational, not simply natural, resources encouraged the TVA to include human and institutional dimensions as well. Other prominent New Deal initiatives included the Public Works Administration (1933) and the Works Progress Administration (1935), agencies that provided loans and grants as well as a reservoir of workers for state and local construction projects.

As part of this development, budgets boomed for the federal water agencies. The River and Harbor Act (1935), for example, seized upon the need to create employment to authorize many major reservoir projects, e.g., Bonneville, Grand Coulee, Central Valley, Fort Peck, and Parker dams, among others. The departments of Interior and Agriculture found opportunities as well: Interior's Reclamation Project Act (1939) authorized multipurpose projects in the West, and Agriculture's Water Facilities Act (1937) called for agricultural water storage and utilization projects in arid and semiarid regions of the country.

The New Deal also spread federal water activity into many new areas. The Public Works Administration gave massive financial aid to state and local agencies for the construction of sewage treatment works. Between 1932 and 1937 the total population served by sewage treatment works increased by 73 percent, and the number of sewage treatment plants increased by one-third (Holmes, 1972, p. 15).

The New Deal was also a time when national resources planning came of age, since it was considered an essential element of economic planning. Although initially directly related to the Depression recovery effort, planning began to assume a life of its own. Under the National Industrial Recovery Act (1933), a National Planning Board was created. Among its first assignments was translating the recommendations of the earlier President's Committee on Water Flow into meaningful action. The committee had targeted ten of the nation's major river basins for priority attention. Incorporating the Public Works Administration's Mississippi Valley Committee as its water planning arm, the board (and its successor agencies) engaged in extensive water resource studies—increasingly on a drainage basin basis. In addition, by presidential directive the board began to assist the Bureau of the Budget in reviewing and evaluating the six-year plans prepared by the federal construction agencies.

By 1939, the board had become the formal planning arm of a new Executive Office of the President. In addition to its role in reviewing and facilitating federal water development efforts, the board began to stimulate planning activities at state and regional levels and to engage in cooperative planning with the states for water resources with the help of Public Works Administration and Works Progress Administration personnel. For the first time, the states were invited into the federal planning process.

Two main themes emerged from the board-sponsored inquiries. The first was a growing commitment to the concept of integrated river basin planning, an approach for which the federal construction agencies were already preconditioned. The other theme was a call for a system of economic analysis to be applied uniformly to all federal water resources projects. Congress encouraged this approach through such legislative enactments as Title I of the Flood Control Act (1936), the statutory genesis of the requirement for benefit-cost analysis. Prodded by the broad concerns of the national planning agency, local and regional water specialists found themselves negotiating the uneven ground of economic analysis, at the same time becoming aware of the connection to social accounting as well. The earlier, ideological principle that development project benefits (with the exception of irrigation) are so widespread and general that specific beneficiaries are unidentifiable began to give way to the concept of repayment of costs to the extent possible. The only exceptions were to be special aid for economically distressed areas and social groups, principles around which distributive politics would rally for decades to come.

Ending of the New Deal and the Wartime Period

By the close of the New Deal period and the advent of World War II, the problem of coordination among federal water resources agencies could not be overlooked anymore. Competition and conflict were rife in the federal water resources effort. The bureaucracies themselves were much to blame, particularly at the geographic interface of east and west in the Missouri and Mississippi river basins, but the move to multipurpose projects had also created functional conflicts among agencies. Congress had not only failed to resolve the problem, but actually contributed to it by overlapping and even conflicting authorizations. Problems had been further exacerbated by the fragmented system of legislative jurisdiction. Virtual client relationships had sprung up between certain agencies, committees of Congress, and special interests. There was also growing conflict between the executive and legislative branches as to who should initiate water projects, apply criteria and standards, and allocate the implementing resources.

Sensing that the handwriting for change was already on the wall, the first remedial approaches were taken by the development agencies themselves. There was obvious conflict between the program objectives of the Corps of Engineers and Interior's Bureau of Reclamation in the Missouri basin, and the arbitrary division of responsibility between headwater and downstream concerns had led to major disagreements between the Corps and the Department of Agriculture as well. The cross-cutting concerns of the Federal Power Commission, and the prospect of other independent regional development agencies like the Tennessee Valley Authority, were threatening to all the federal construction agencies. Major General Julian Schley, the Chief of Engineers, actually testified in Congress against the president's 1937 proposal to set up TVA-like entities for each of the nation's seven major river valleys. He said that it would interfere with the Corps's traditional relationship to Congress and to individual congressmen (Holmes, 1972, p. 22).

Under the precedent of a tripartite agreement signed by the departments of War, Interior, and Agriculture in August 1939, the water agencies were spurred to form a voluntary organization, the Federal Inter-Agency River Basin Committee (FIARBC), when the National Resources Planning Board was abolished in 1943 by a vindictive Congress. Agriculture, Interior, War, and the Federal Power Commission were the charter members of FIARBC, with Labor and the Federal Security Agency (the first federal water pollution control agency) added later.

With membership drawn largely from the subcabinet level, the committee was momentarily out of the growing conflict over water policy between the president and Congress. It was able to form technical and special subcommittees to work on common problems (e.g., the "Green Book" of suggested project justification procedures released in 1950). Even more importantly, it could begin to create a working network of regional, interagency committees for specific river basins, thereby extending the principle of coordination to the field level. Chaired by the federal agencies, often on a rotating basis, these interagency committees began to follow the practice of including state agency representatives as participating advisors. Although competition was still keen among the development agencies, the commonly accepted necessity of integrated basin surveys, and the personal rapport that began to develop among federal and state officials, started to leave its mark on the water policy scene.

Throughout this period, the attitude of Congress toward coordination was ambivalent. On the one hand, various legislative enactments (e.g., the Federal Water Power Act of 1920, the interagency consultation provisions of the Flood Control Act of 1936, and the Pick-Sloan compromise for the Missouri basin contained in the Flood Control Act of 1944) moved the federal water agencies toward a resolution of their differences. On the other hand, it was generally advantageous for Congress to maintain a divided set of responsibilities within the executive branch. Even after the Legislative Reorganization Act of 1946, Congress kept three separate standing committees for water projects, one for each of the principal federal construction agencies.

The Policy Commissions: First Round, 1950–1961

In the aftermath of the concerted national effort of World War II, a fragmented and bloated federal bureaucracy was no longer tolerable. With bipartisan support from Congress, the first U.S. Commission on the Reorganization of the Executive Branch of the Government (the Hoover Commission) was created. Although its primary objective was to find ways of saving money and achieving efficiencies throughout the government, the commission gave special attention to the problem of overlapping and uncoordinated water agencies. In its first report (1949), the Hoover Commission recommended a consolidation of the Corps's civil works program and the Bureau of Reclamation within the Department of the Interior as a new Water Development and Use Service. It recommended an independent Board of Review for water projects in the executive, and a system of drainage area commissions, with federal and state representatives, for coordination and advisory purposes. Its second report (1955) reached even further—it recommended a consolidation of the Department of Agriculture's upstream and the Corps's downstream flood control

programs; a new Water Resources Board under the president to recommend policy and coordinate planning; a strengthened process of
project evaluation by the Bureau of the Budget; and federal action
(through the new Water Resources Board) to set up interagency river
basin commissions. In a more stringent political and budgetary environment, coordination had moved beyond the status of a desired objective to become a necessity.

Then began a remarkable series of water resources inquiries, not
just to determine the direction of federal policy but also to gather the
political support necessary to bring about the required reforms.

The first of these was the President's Water Resources Policy
Commission, established by President Truman's executive order in
January 1950. Chaired by Morris L. Cooke, a nationally respected
civil engineer and the former chairman of the Public Works Administration's Mississippi Valley Committee, the commission included
six other water and public works professionals. The charge to the
commission from President Truman highlighted four aspects of water
policy: the extent and character of federal government participation;
an appraisal of program priorities from the standpoint of human and
social need; criteria and standards for evaluating the feasibility of
projects; and new or modified legislation needed.

The commission cast its net broadly, and by December 1950 the
assessment had been completed. It recommended that:

• Congress should set up major river basin commissions under independent
chairmen appointed by the president.
• A Board of Review should be formed to uniformly consider programs,
establish procedures for evaluation of direct and secondary benefits and costs,
and insure that proposed projects were considered not in isolation but as
parts of overall multiple-purpose basin programs.
• Geological and hydrological data should be systematically gathered to facilitate annual estimates of water use or need versus available surface and
groundwater supply.
• The policy recommendations should be incorporated into a single statute,
applicable to all federal water resources activities and agencies.

Despite setting a new standard of comprehensiveness, in retro-
spect the Cooke Commission report was deficient in several respects.
There was a singular lack of emphasis upon the needs of recreation,
fish, and wildlife. Municipal and industrial water supply needs were
given short shrift, and the commission suggested waiting 10 years
before further federal intervention in water pollution control. Two

major issues were addressed but not resolved in the commission's final report. One was irrigation program costs; the other the need for an organizational rearrangement of water resources agencies to implement the report's findings. The cornerstone of the latter recommendation—a single statute applicable to all federal water resources activities and agencies—was drafted but never implemented. The Cooke Commission report nevertheless stimulated attention to policy matters in a number of quarters. The Bureau of the Budget, for example, was inspired to issue Circular A-47 during the final days of the Truman administration (December 1952) setting forth the standards to be used by all agencies in evaluating water resources projects.

The Eisenhower administration established an interagency, cabinet-level Presidential Advisory Committee on Water Resources Policy in 1955 to consider the recommendations of both the Cooke Commission and the Hoover Commission. Five years after the Cooke Commission report, the primary objective of water resources planning seemed to have shifted away from regional development to the apportionment of an increasingly scarce resource, a concern prompted by the unprecedented increases in industrial water usage during the early part of the 1950s.

Congress, too, began to pay more attention to federal water policy. Congressman Robert N. Jones, chairman of a special subcommittee to study civil works of the House Public Works Committee, recommended that Congress issue policy statements determining the federal role in water resources development, the place of local and state interests in federal development, uniform standards for economic justification of projects by all executive agencies, uniform standards for allocation of costs in multipurpose projects, and uniform criteria for the sale of products to recover such costs. And on the Senate side, prompted by increasingly divisive positions on water projects by the Republican administration and the Democratic Congress, Senate Resolution 48 was adopted in April 1959 establishing a Senate Select Committee on National Water Resources. The committee consisted of 17 senators, chaired by Senator Robert Kerr (Oklahoma).

The Kerr Committee presented its report to the Senate in 1961. Among its principal recommendations were many echoes from the earlier Cooke Commission and other policy reports—the need to avert water shortages likely to be produced by anticipated economic growth, an emphasis upon the maximization of national welfare objectives (e.g., jobs and income), and a reliance on technical measures to accomplish the above (reflecting a new fascination with space age technology). The Kerr Committee asked the federal government to prepare and keep up-to-date plans for comprehensive development and management within all of the major water resources regions of the country. It stressed that greater efficiency in water use and development must be achieved to avoid problems and meet the needs for the future. The most important step toward advancing these recommendations and solving the nation's water resources problems, the committee concluded, was to increase public awareness of the nation's water resources.

Unlike the Cooke Commission, the Kerr Committee did not propose legislation to implement its recommendations. The Senate members, senior in their respective standing committees, were expected to see to that need. The 1960 elections proved provident, for on July 13, 1961, President John F. Kennedy recommended legislation embodying many of the committee's proposals. The committee findings on coordination and assessment, river basin commissions, and the stimulation of state programs emerged as titles of the Water Resources Planning Act of 1965. The Kerr Committee's research recommendations were incorporated in the Water Resources Research Act of 1964. President Kennedy, however, followed the "old" conservationist line of the Kerr Committee report even to the extent of rescinding Circular A-47 in favor of Senate Document 97, which made it easier to justify the old-style water projects.

The Policy Commissions: Second Round, 1968–1980

Less than a decade later, yet a third major assessment of federal water policy was to be undertaken, this time spurred by dissension over the use of waters in the Colorado River basin. The massive Central Arizona Project was believed to be capable of diverting more than the long-term average supply of water available. It was the Bureau of the Budget's opinion that an objective study of the water problems of the entire United States was needed in order to place the Colorado basin proposals in proper perspective.

In response, a seven-member National Water Commission was created by Congress in 1968 with former Interior under secretary and Consolidated Edison board chairman Charles F. Luce as chairman and six other independent members appointed by the president. After five years of studies and research, the commission's final report was transmitted to President Nixon and Congress in June 1973. More than 200 recommendations were offered for improving national water policy.

In a major departure from previous reports, the commission declared that future demands for water are responsive to water policy and therefore that plans should be made to design the future rather than accept it as given. It forecast a marked shift of priorities away from water development toward the preservation and enhancement of water quality, and it asserted the need to tie water resources planning more closely to land planning. The commission favored greater use of economic approaches to reduce water losses, increase efficiencies, and otherwise advance water conservation, and extended that concept to embrace the beneficiary-pays principle. It called for a reexamination of laws and legal institutions governing water resources as long overdue, and tried to define the role of the federal government in water resources matters. In general, the commission concluded, the front-line actor should be not the highest level of government but "the level of government nearest to the problem with the capacities required to represent all the interests and resolve the matter in timely and equitable fashion."

Through the report issued by the National Water Commission, the federal government finally came to grips with issues that had been obvious to many other levels of government for some time, in particular the problems that flowed from the fact that the United States had evolved into a predominantly urban nation. One could argue whether environmental pollution was objectively worse by the early sixties than it was a century earlier, but this was irrelevant to the actual views that the large and increasingly affluent urban populations held regarding their environment. And things were not good. A series of environmental accidents (some would call them disasters) happened during the 1960s that helped foster the impression that everything was going to hell. Massive fish kills on the Mississippi during 1963–1964 served to underline the thesis of Rachel Carson's 1962 book *Silent Spring,* even though the kills were eventually traced to the effluent of one single industrial plant near Memphis, Tennessee, and hence had little to do with Carson's theories about widespread use of pesticides. (Carson's fears were realized in the 1968 discovery of very high levels of DDT and other pesticides in some species of commercial and sport fish in Lake Michigan.)

Nevertheless, environmental events quite properly frightened many people and, more importantly, called into question the efficacy of the cumbersome federal regulation of water pollution. In 1969 there was a blowout of an oil platform operated by a federal lessee on the continental shelf near Santa Barbara, California, that created an oil slick covering hundreds of square miles and fouled 13 miles of shoreline, causing great damage to the fish and wildlife and also to the tourist industry. This was followed by the discovery of potentially dangerous concentrations of methylmercury in U.S. streams. The chlorine industry was estimated to be dumping a million pounds of the substance each year into U.S. and Canadian waters. In response to the public health threat, industry moved quickly and by the end of 1970 had reduced the effluent to less than 15,000 pounds per year: a 98 percent reduction in one year! The list of reckless dumping of wastes or filling of wetlands for airports and other construction was large and growing.

At the same time demographics led to a slow change in the composition of the key resource committees in Congress. The concerns of the new congressmen and their constituents were for water pollution control, municipal water supply, outdoor recreation, and the protection of scenic and unique sites as tourist attractions. The Johnson administration can be considered the transition from the development orientation of the Kerr Committee to the quality orientation of the Luce Commission and afterward. Despite President Johnson's 1966 message to Congress on the preservation of America's natural heritage, in which he stressed that pollution (particularly water pollution) was the most urgent and costly natural resource problem, he actually had no intention of spending money to fix it (Holmes, 1979, p. 98). Congress, however, responding to its constituents, had different ideas about water pollution control—money. The Clean Waters Restoration Act of 1966 established an entirely new scale of federal funding for water quality. The funding level for the construction grants program was raised from $150 to $450 million per year for fiscal 1968, $700 million for 1969, $1 billion for 1970, and $1.25 billion for 1971. These were unprecedented amounts, equal to the size of the Corps of Engineers program. The authorizations were, however, not fully appropriated during the last two years of the Johnson administration due to the cost of the Vietnam War.

Despite the existence of a large environmental bloc in Congress and the publicity given to environmental disasters, natural resources and the environment did not feature in the 1968 Nixon-Humphrey presidential election campaign. Nonetheless President Nixon saw the environment as an issue strongly supported by the public and one that his administration could exploit. Upon taking office he was immediately faced with the Santa Barbara oil spill and did a creditable job in dealing with it. The coincidence of a new president taking office 20 years later and facing a similar oil spill should not go unnoticed— the anemic Bush response to the Alaskan oil spill of 1989 makes the vigorous Nixon response of 1969 seem all the more remarkable. Nixon was, however, unable to sustain this early success because of the continuing financial drain of the Vietnam War.

Nixon was originally ahead of Congress on environmental issues, including the creation of an Environmental Financing Agency. Its purpose was to help communities finance their share of treatment plant construction costs, allocate funds to areas where the greatest improvements in water quality would be achieved, extend federal jurisdiction to all navigable surface waters and interstate groundwaters, ensure more efficient enforcement procedures, and provide court-imposed fines for polluting of up to $10,000 per day.

In 1969 President Nixon appointed a Task Force on Resources and the Environment under the president of the Conservation Foundation, Russell Train, established a Cabinet Committee on the Environment to advise him, and signed into law (January 1, 1970) the National Environmental Policy Act of 1969 (NEPA), which included the establishment of the Council on Environmental Quality. The administration diligently enforced the NEPA against strong opposition from the traditional federal water and other resource agencies, and in December 1970 established on its own initiative the Environmental Protection Agency.

The administration based its arguments for the EPA on the need to have the environment considered as a single interrelated system; it brought together from various other agencies the pollution control programs for water, air, solid wastes, pesticides, and radiation. Yet only four years earlier the Johnson administration had used similar arguments to move the Federal Water Pollution Control Administration from the Department of Health, Education, and Welfare to the Department of the Interior on the basis of needing to get the water programs under one agency. This splitting of water issues between agencies still plagues federal water policy to this day.

Some analysts (Holmes, 1979, p. 108) speculate that the removal of the federal water quality agency from the Department of the Interior to the EPA, which did not have full membership status on the Water Resources Council, led to a serious shortcoming in the council. Water quality was not fully integrated into its discussions. The EPA appeared to reject the idea that interagency river basin planning could lead to optimum use of water for all purposes (including water quality), as well as the notion that water pollution control and water conservation had the same basic aim—based perhaps on the events of the 1960s and the seeming lack of success in the earlier federal efforts in water quality management.

In any event, the 1972 Clean Water Act with $18 billion for the construction grant program, its emphasis on the ecology of the streams rather than their use for society, and its rigid adherence to technology-based standards left little room for discussion with the traditional water agencies. The EPA was essentially off on its own, with the largest of the federal programs and the smallest technical manpower to administer them.

The National Commission on Water Quality.   The ink was hardly dry on the National Water Commission report before another independent commission launched an inquiry into another aspect of water resources policy. Mandated by the Federal Water Pollution Control Act of 1972, the National Commission on Water Quality began its work during 1973 with five presidential appointees and five each from the House and Senate, under the chairmanship of the vice-president. The commission's assignment was to consider the midcourse corrections necessary to the stringent requirements of the 1972 Federal Water Pollution Control Act, measures that had mandated fishable-swimmable waters nationally by 1977, and zero discharges of pollutants into navigable waters by 1985.

Driven by technology-based standards ("best practicable" technology—BPT—by 1977, and "best available" technology—BAT—by 1983), the cleanup effort proved more complex than envisioned. Sensitive to these realities, but equally sensitive to the potent political forces opposed to any relaxation of standards, the commission recommended that the 1977 and 1983 shortfalls should be met by deadline extensions and waivers on a case by case basis, not by a general extension of the deadlines. For the 1985 deadline, however, the commission favored a five- to ten-year blanket extension with a similar commission established in 1985 to weigh progress and recommend further revisions. The commission urged retention of the fishable-swimmable goals (pushed back five years to 1982), but favored a substantial softening of the 1985 zero discharge goals.

A host of recommendations was issued with regard to specific pollutants and pollution sources. Among them was a generic warning that non-point sources, and the elimination of toxic pollutants, were topics of growing urgency. And on the matter of administration and funding, the commission found the Environmental Protection Agency's administrative capacity to be severely overtaxed. The required tasks of issuing rules, regulations, permits, effluent guidelines, and limitations were simply monumental. Shortfalls in funding for the large construction grant program, and slow progress in obligating even the funds available, had led to significant delays in achieving the 1977 requirements. Little or no effective planning had been done.

The commission observed that many of the administrative difficulties could be relieved if administrative and regulatory functions were turned over to the states. But the most critical element of the program was construction grant funding. Not only should the 75 percent federal cost share be retained, the commission reported, but the program should be level-funded for five to ten years at a rate of $5–10 billion per year to stabilize the planning of municipal treatment facilities, the key feature of the entire national water pollution control effort.

President Carter's policy initiatives.   Soon after the National Commission on Water Quality's final report to Congress in March 1976, a new administration was poised to take over in Washington. President Jimmy Carter had been exposed to the failings of federal water policy from his days as governor of Georgia. When a suggested list of water projects to be canceled was brought to his attention early in his administration, the Department of the Interior was instructed to make a comprehensive study of the initiatives needed in water policy reform. Utilizing 13 separate working groups and many water experts from outside the government, the Carter task force recommended revisions in several principal areas of water policy in its report of December 6, 1979.

The President's Water Resources Policy Study Task Force, as the group was known, is best known for its recommendations in three aspects of policy: its advocacy of a role for the states in federal project decisions; its support for cost sharing and pricing reforms, and its recommendation that the Water Resources Council be responsible for the application of evaluation standards to all federal water projects. Under the new ground rules proposed, many of the western water projects would no longer qualify. Both the prospect and the reality of the new policies were in sharp contrast to the historic prerogatives of Congress in determining project authorizations.

The protest fell most heavily on the Water Resources Council, an interagency planning and coordinating body established under the Water Resources Planning Act of 1965. Among its other legislated functions, the council was required to prepare regular assessments of the nation's water resources. The council's second national assessment was issued in 1978, coincident with the Carter water task force deliberations. Three changes in national administration had occurred within the 10 years since the first assessment (1968), and there had been a marked change in operational emphasis to favor data collection and compilation over analysis and evaluation. Despite the widely held view that shortcomings in the assessment process rendered the report of questionable utility, the council's findings attracted attention.

The most startling observation made in the second assessment was that water withdrawals (the amount actually drawn from the ground or surface waters), rather than increasing, would decline by 10 percent by the year 2000 as the result of diminishing industrial needs. Within the same time period, however, there would be a substantial increase (27 percent) in water consumption (that amount withdrawn and not returned to the relevant hydrological cycle within the same water year). According to the council's assessment, maintenance of urban domestic and commercial supplies would become the key challenge for water resources managers in the upcoming years, followed closely by the need to meet growing environmental and recreational demands. The latter would require careful research if instream ecological values and flow requirements were to be considered fully in planning and decision making.

Ten critical problems were described, including:

• Inadequate surface water supplies would become apparent in 17 subregions.
• Overdraft of groundwater would be an extensive problem in eight subregions—notably the high plains area from Texas to Nebraska—and a moderate problem in 30 others.

• Pollution of surface water by point and non-point sources would occur in most of the country.
• Damage from flooding could amount to $4.3 billion annually if floodplain management efforts and regulation were not accelerated.

At the time of data collection (1975), areal soil losses from croplands were already amounting to nine tons per acre per year, and from forest and pasture lands an additional one ton per acre. The council found that six million acres of wetlands had been lost to dredging and dredge spoil disposal from 1955 to 1975, and another eleven million acres were likely to be converted to agricultural croplands by the year 2000. And the degradation of the nation's bays, estuaries, and coastal waters by domestic and industrial wastes was found to be creating significant conflicts with recreational, wildlife, and fishing interests.

Under the provisions of the Water Resources Planning Act of 1965, the Water Resources Council was mandated to prepare assessments "biennially, or at such less frequent intervals as the Council may determine." The council was administratively terminated in 1981 by the incoming Reagan administration, and its formal assessments replaced to some degree by the more modest data reporting of the U.S. Geological Survey. Nevertheless, the statutory provisions of the 1965 act have never been repealed and are still law.

The Reagan Years

With the memories of the confrontation with the Carter administration fresh in mind and a new administration now in office, the Subcommittee on Water Resources of the Senate Committee on Environment and Public Works held two days of oversight hearings in late 1981 and early 1982. Two Reagan administration witnesses (William Gianelli, Assistant Secretary of the Army for Civil Works, and Dr. Garrey Carruthers, Assistant Secretary of the Department of the Interior for Water and Science) assured the subcommittee that the institutional vestiges of the Carter administration—the Water Resources Council, the seven river basins commissions, and the state grants program—were being swept out in favor of a cabinet-level Council on Natural Resources and a new Office of Water Policy in the Department of the Interior. In the absence of the Water Resources Council and regional river basin commissions, members of the subcommittee were uncertain how the administration intended to establish priorities, fulfill past commitments, remove impediments to project development, be equitable to regions, and involve the states. It developed that not all of the previous Carter initiatives had been found wanting, for the new administration planned to install government-wide procedures on project evaluation—termed guidelines rather than standards—that would usher in a new era of partnership with the states and local project sponsors. This was expected to clear the way for the first omnibus water projects bill in more than a decade (an interval caused by gridlock between Congress and the president—first Carter, then Reagan—until Reagan got what he wanted).

True to their word, Reagan administration officials began working with federal water agencies and the substantive committees of Congress on a new omnibus water bill, and in December 1986 the Water Resources Development Act became law. It authorized more than 270 water projects, numerous studies, project modifications, and miscellaneous projects, at an ultimate federal cost of $16.5 billion. But the policy provisions of the act also portended substantial changes in the way the federal government would undertake civil works in the foreseeable future. Among the most significant were the requirements for an equal financial contribution by a nonfederal sponsor, cost-shared planning, an increased nonfederal cost share for project construction, and up-front nonfederal financing during actual construction. Also the fuel tax to pay for the Waterways Maintenance Trust Fund was to be raised gradually from 10¢ a gallon to 20¢ a gallon by 1994 and a separate Harbor Maintenance Trust Fund for dredging harbors was set up.

In a remarkable display of bargaining and consensus building, there was something for everyone in the final act. Even the environmentalists, the legendary foes of water projects, were able to accept the Water Resources Development Act, for the beneficiary-pays principle promised to scale down projects and weed out the most unsatisfactory at an early stage. Much of the debate on the legislation occurred backstage, a product of the active involvement of the Office of Management and Budget with the substantive and appropriations committees of Congress.

Though signing the Water Resources Development Act, President Reagan vetoed the 1986 amendments to the Clean Water Act because of the five-year $18 billion cost. The 100th Congress reintroduced and passed the bill as the first item of business and subsequently overrode the president's second veto. Unlike that of the Water Resources Development Act, the money authorized by this bill was not dependent on the availability of matching funds—since the municipalities were mandated to move forward on the construction of facilities, increased local participation would just broaden the numbers of municipalities receiving the funds.

By the start of the Reagan years, the third element of federal water policy—regulation—had also become intolerable to many water users. Regulation had been ushered in by the emergence of the national environmental movement in the 1960s. Shocked by a series of revelations about particular instances of environmental deterioration, the public demanded tough action. The relative affluence of the period made it unlikely that the economic effects would be severe, and the growth of leisure-time activities like outdoor recreation gave Americans the chance to enjoy the benefits of regulation firsthand.

With a benign view of government still prevailing, the question was usually not whether regulation was the solution but rather how much regulation was necessary. The case for water-related regulation was also helped measurably by the growing public disenchantment with traditional water projects. In the 1960s and 1970s the Corps of Engineers, for example, had encountered stiff resistance to its plans for a Cross-Florida Barge Canal and its proposed system of reservoirs in the upper reaches of the Potomac basin (which were justified primarily for low flow augmentation and water pollution control), and Interior was halted by opposition to the multipurpose reclamation dams planned for the splendid scenery of the Grand Canyon. And concern for a six-inch fish—the snail darter—was destined to topple, if only temporarily, TVA's proposed Tellico Dam on the Little Tennessee River (Tellico was reauthorized by a rider to the annual public works appropriation bill by Representative John Duncan in 1979, and "in forty-two seconds the citizen's work of sixteen years was reversed" [Plater, 1982, p. 783]).

Fueled by broad environmental statutes like the National Environmental Policy Act of 1969, and backed by the new army of environmentalists, the fever for governmental regulation had appeared to have no limits. When President Nixon established the Environmental Protection Agency in December 1969, there was at last a federal agency to enforce the rules. But by the early 1980s, even hard-nosed enforcement advocates like former EPA assistant administrator John Quarles admitted that Americans may have committed themselves to goals and programs that would turn out to be unrealistic.

The Bush Years

Campaigning for the presidency in 1988 from a boat in heavily polluted Boston Harbor, George Bush declared himself to be an environmentalist and pledged to become the "environmental president." The National Wetlands Policy Forum had recommended in November 1988 that a policy of "no overall net loss" of wetlands be adopted by the federal government. The incoming Bush administration eagerly embraced this idea and also that of wetlands restoration and creation where possible. The Corps of Engineers and the EPA jointly attempted to work out procedures and regulations to develop the concepts. In order to follow a policy of "no net loss" it was important to have an adequate definition of what is and what is not a wetland. A major dispute arose with the release in 1989 of a revised version of the 1987 wetlands delineation manual. The 1989 version was more restrictive and led to much more land being defined as regulated wetlands. The ensuing uproar led Vice-President Quayle's Competitiveness Council to become embroiled in the debate and the Corps of Engineers and the EPA to try to agree on an improved delineation. On August 9, 1991, the Bush administration presented a wetlands delineation manual with a much more narrowly restricted definition of what constituted a wetland; so narrow that even the administration's own scientists claimed that it would effectively remove protection from as much as one-half of the wetlands defined under the 1989 manual. (Of course it is not safe to assume that the areas designated under the 1989 manual were themselves the correct total. Maybe the right delineation lies somewhere between the 1987 and the 1989 manuals.)

Congress's anger at this confusion is manifest by its explicit embargo in the Energy and Water Appropriations Act passed at the beginning of October 1992, which ignored the 1991 version of the manual and prohibited the Corps of Engineers from using the 1989 version for wetland delineation. It also appropriated $400,000 to fund a definitive study of wetland delineation by the National Academy of Sciences. During the 102nd Congress more than 20 wetland-specific bills were introduced; all died during the Congress. One can only expect

that in the 103rd Congress, along with the reauthorization of the Clean
Water Act, some form of wetlands-specific act will also be passed.

Despite some success in other phases of the environment, notably air pollution, the Bush presidency achieved little in the water area.
Despite great promises about wetlands, the review of wetlands regulation got bogged down in an endless debate in the Domestic Policy
Council of the White House and was subject to sniping by the vice-
president and his Competitiveness Council. Largely due to the trouble with wetlands, the revisions of the Clean Water Act, supposedly
to be taken up during the 102nd Congress, were not pushed by the
White House or Congress. After four years little progress had been
made on water policy.

The 1986 Water Resources Development Act signaled a return
to a two-year cycle of water bills. A Water Resources Development
Act was passed in 1990 that authorized 26 new water projects, and
just prior to the 1992 presidential election, in a flurry of last-minute
activity, Congress passed and the president signed into law a number
of water or water-related bills: on October 2, PL 102-327, Fiscal Year
1993 Energy and Water Appropriations; on October 5, PL 102-381,
Fiscal Year 1993 Department of Interior Appropriations; on October
6, PL 102-389, Fiscal Year 1993 Appropriations for Small and Independent Agencies; on October 30, PL 102-575, Reclamation Projects
Authorization and Adjustment Act of 1992; and on October 31, PL
102-580, Water Resources Development Act of 1992. This last bill,
the omnibus water bill with $2.2 billion of federal project money,
includes 23 new flood control, navigation, and other projects (including $519 million to restore the Kissimmee River in Florida to its orig-
inal meandering course).

From a policy point of view the Reclamation Projects Authorization and Adjustment Act is the most interesting. Most importantly,
it rearranges the western water supply picture, particularly in California. Key provisions of the bill will decrease the water flowing to
irrigation in California's Central Valley Project by as much as 1.2
million acre-feet, diverting it for fish and wildlife uses. This amounts
to an unprecedented 20 percent reduction in water supplied to agriculture. Other provisions of the bill include $992 million for the com-
pletion of the Central Utah Project (the very last of the federal water
juggernauts). It also has provisions allowing farmers to sell federal
water to urban areas. Title XXX of the act, the Western Water Policy

Review Act of 1992, sets up an advisory commission to review the problems of "western water." Another provision of the act requires that the operation of Glen Canyon Dam be changed away from maximizing power generation during peak flows to protecting, improving, and mitigating adverse impact to the Grand Canyon's riparian habitat.

Signing this bill required some fortitude on George Bush's part since many of the bill's opponents were staunch Republican supporters in key states. Environmentalists were delighted with the bill, since to them it represented another block cemented into the federal policy of moving away from wasteful water development to more ecologically responsible stewardship of the resource.

One very important piece of legislation also signed by the president (October 24, 1992) that is often overlooked by water policy commentators is the Energy Policy Act of 1992. Ostensibly concerned with energy conservation, this piece of legislation has potentially major direct impacts on water supply and wastewater disposal for the entire country. Tucked away in the fine print of Section 325 are national water efficiency standards for all new faucets, shower heads, and flush toilets for residential use manufactured after January 1, 1994. The standards set by the legislation are 2.5 gallons per minute for faucets and 1.6 gallons per flush for flush toilets. The standards for shower heads are to be set within 12 months.

As can be seen from table 2.3, these three fixtures constitute, on average, almost two-thirds of domestic indoor water use. The implementation of the new standards should easily reduce domestic indoor water use by up to one-third. As the new fixtures are phased in with the normal replacement of household plumbing fixtures, we can expect to see a continual decline in the demand for indoor water use due to this legislation alone. In addition to the direct impact due to the national water efficiency standards, there will be large indirect impacts on water due to the energy conservation provisions of the act. It is well known that many of the uses of energy in industry and commerce involve heating and cooling water. Usually any improvement of the efficiencies in these processes involves recycling the water to recover the heat and, hence, leads to significant reductions in water withdrawals. These direct and indirect effects will continue to exert a downward pressure on the already declining demands for water.

Federal Water Policy in Retrospect

This abbreviated look at the history of water resources policy[2] indicates that water projects continue to represent one of the few vestiges of old-style power remaining in Congress. The water lords of yesterday still exert undue influence over certain federal programs (e.g., irrigation), and the antagonisms of the past—often more symbolic than actual—still get in the way of progress. Although pervasive new constituencies such as environmental interest groups have arisen since the early 1970s, the process of bargaining remains much as it was in the last century. Reforms must still be sold to strategic players in an atmosphere of conflict, logrolling, and potential blackmail. There are still problems in state-federal relations, and divisive conflict remains between elements of the federal bureaucracy. There is still no substitute for a personally committed chief executive, directly accessible by cabinet officers and policy staff, as the most efficient agent for change.

This overview also shows that policy change in water resources, as in other aspects of national policy, is often a step behind the evolution of social goals. While the problems to be addressed are invariably long-term in nature and public and expert views are always in movement, political change occurs only at electoral intervals of two, four, or six years. This time lapse weakens the link between planning and implementation. It also suggests that the scale and complexity of water problems across the nation have outgrown the institutional capabilities at hand.

What, then, is the outlook for the foreseeable future?

First, it is clear that federal water policy does not exist in and of itself but merely reflects the times. It is just one of many policy tools used to address societal problems and needs and therefore cannot be viewed outside of its broader economic, social, and political context.

Second, there are times when policy change is politically possible and times when it is not. Moments of consensus are exceedingly rare. Entrenched special interests—the forces of precedent, habit, and agency turf, plus sheer inertia—mean that changes in federal water policy do not come about readily or gracefully. Years may be required to bring about even the simplest reforms.

Third, a pendulum effect is seen in much of the history of federal water policy. A topic will be hot at one time and ignored at the next. Development concerns may dominate the policy agenda for a while, followed by a period of checks and balances on those same development processes. An apparent commitment to regulation may be superseded by concerted and successful resistance on the part of those regulated. At some moments, the "harder" aspects of water management may prevail—dams, reservoirs, waterways, improvements, and works for irrigation and hydroelectric power. At others, the "softer" aspects—planning, coordination, research, data, and environmental concerns—will dominate the policy agenda.

Fourth, there are a variety of engines that trigger change in federal water policy. They may be cataclysmic national events, such as war, economic depression, or natural disaster, or a clear-cut national commitment to a goal, such as the settlement of the West, the need for national or regional economic development, or the restoration of environmental quality. Water is often brought along as part of a larger agenda.

Finally, there are individual policymakers who make a difference. President Franklin D. Roosevelt exemplified the role of the strong chief executive; Senator Robert Kerr personified the powerful, knowledgeable, and respected legislative leader. Inside the federal establishment, particular areas of competence or influence may lead to significant policy change. In President Theodore Roosevelt's administration, Forest Service Chief Gifford Pinchot exerted unusual influence on the nation's emerging natural resources agenda. In the Reagan years, Interior Secretary James G. Watt and Army Assistant Secretary William Gianelli had the confidence of the president and made their views strongly felt. Individuals and organizations outside the federal government can be equally persuasive, for example an influential governor (Bruce Babbitt of Arizona or Jimmy Carter of Georgia), responsible interest groups (the Clean Water Coalition or the National Water Alliance), or respected professionals (the late Abel Wolman, a sanitary engineer from Johns Hopkins University).

Many observers have lamented the absence of a national water policy. Yet history indicates that we do have a national water policy, however incrementally designed, consisting of separate pieces, projects, and actions, ultimately responsive to specific opportunities or problems, and surprisingly free from partisan politics. The question must be asked whether the essentially ad hoc approach to policy formation can and should be improved. Over the years, a number of cross-cutting themes have been identified that indeed seem worthy and ready for remedial attention.

Water History: A Nonfederal Perspective

Before leaving the history of the federal role in water policy, it should be made clear that this is only one part of the picture; many interesting things have taken place outside the federal purview. The major difference between federal and nonfederal concerns is that for over 100 years the federal government ignored issues of water quality and public health—*the* major concerns of the nonfederal entities. From 1776 until 1900 cities and localities could have benefited greatly from federal help in combating waterborne diseases, but received none. The greatest irony is that by 1970, when most of the local communities were finally well on their way to dealing effectively with these issues, the federal government started to push hard on human health as the remaining unsolved water problem. In fact, the drinking water quality is better now than at any point during the past 200 years for most citizens, and certainly for urban dwellers.

Appendix 3 gives a rough chronology of U.S. water history from a nonfederal perspective, relating the implementation of local water policy to the concerns of the times and the changes in technology that structured the policy responses. This chronology is eclectic in that it treats only a few of the cities and regions, whereas the actual events were reproduced widely over the nation. As a result, Boston, Philadelphia, and New York appear more frequently than if the chronology were exhaustive.[3]

Over the 350 years covered in appendix 3, water's role in economic development was a much lesser concern than the protection of human health. This is the reverse of the emphasis of federal policy of the time (appendix 1), perhaps based on the belief that while economic issues might take care of themselves, health was too important an issue to deal with at the federal level and must be left to local government.

The role of the private sector in the development of water policy is one of several areas subject to trends and cycles occasioned by changes in political philosophy. Initially private development was overwhelmingly the approach chosen; later, with the growth of large cities with strong governments, local government took over most of the private water supply and sewer companies; recently, since the start of the Reagan era, there has been pressure to have the private sector assume many of these responsibilities. In Britain, perhaps as a consequence of the developments in the United States, Prime Minister Margaret Thatcher put the Regional Water Boards up for auction. The market response has been quite favorable, with some of the largest utilities being sold to large French private water utilities.

Another area subject to changes in policy is waterborne diseases, based on the development of theories of how they are transmitted. An early idea was that disease arose out of the "miasma" and odors emanating from untreated human wastes. Based on this theory, the solution, at a time when the nation had a relatively small population, was to transmit the wastes to a remote location away from the people. Consequently, much early work was devoted to building sewers and simply removing the wastes. Later, when the germ theory of disease was better understood, the solution was to treat the water to kill or remove the dangerous pathogens. More recently, with perceptions that toxic and potentially carcinogenic chemicals in the drinking water might cause cancer at a zero tolerance level, communities now are looking for solutions that either switch the source of supply to some uncontaminated area (equivalent to moving the waste out) or remove all but the remotest amounts of chemical from the water (equivalent to finding a new source of supply). The current emphasis, which is encouraged by the federal agencies, is without strong epidemiological evidence of actual human health risks.

Water as a Resource: What Makes It Different?

Water as a Natural Resource

Despite the growing complexity of modern life, and despite the increasing urbanization of the world's population and its alienation from nature, we still live in a world very heavily influenced by the availability and management of natural resources. Like Japan, or on an even narrower base Holland, countries can be rich despite relatively poor natural resource endowments if they concentrate on the trade and transformation of natural resources; but, historically, this is the exception rather than the rule. The United States is one of the few nations that possess a vast array of natural resources—only Canada, Russia, and Brazil have similar endowments.

To place water and water policies in the broader context of natural resources, an understanding of several definitions and philosophical theories underlying government action involving these resources is first required. Most practical people tend to forget that they are constrained in their everyday actions by the mad scribblings of some long-dead intellectuals. Once in a while it pays to remind ourselves of how rooted in theory many of our policies really are.

As opposed to "man-made" resources, Howe points out that "traditional usage confines the term [natural resources] to naturally occurring resources and systems that are useful to humans or could be under plausible technological, economic, and social circumstances" (Howe, 1979, p. 1). Thus, resources are only resources when they are useful to people—when their existence or possession or use is of *value* to some human beings, where "value" includes aesthetic and spiritual values. At first glance this may appear as human chauvinism, but it does not imply that we should ignore the flora and fauna of the natural universe. For a variety of aesthetic, cultural, medical, and ethical reasons, some people care a lot about those parts of the living and nonliving environment. They are thus assigned a value and are resources.

Ciriacy-Wantrup suggested that natural resources and their management—he called it "conservation"—are a problem in human ecology, and that all resources lie in one of three categories: natural, cultural, and human (Ciriacy-Wantrup, 1963). His approach is attractive because of a more than simple coincidence of these categories with the factors of production commonly used in economic analysis—land, capital, and labor.

Natural resources may also be split into nonrenewable or stock resources and renewable or flow resources. Both of these can be further categorized as resources that are easily held as private property, common pool resources, and fugitive resources. The distinction between renewable and nonrenewable natural resources is prominent in our current literature and outlook, often as the vehicle for expressing anxiety about the exhaustion of nonrenewable resources. However, as we shall see, it is not always so easy to determine whether a resource is renewable or not. At the extremes it is easy to decide—a coal seam has a definite finite life that ends if it is continually mined and burned. The survivability of a fishery, however, depends on whether it is managed for short-term profit or for long-term sustained yield. There are obviously many choices to be made in natural resources policy, and the range of choices becomes more complex as one progresses from obvious cases of nonrenewability to situations in which a policy choice helps define the resource itself. For example, the emphasis on particular chemical contaminants in the 1986 Safe Drinking Water Act may lead many communities to abandon their current surface water sources for new groundwater sources. For those communities the resource base will shrink, because they feel pressure to abandon a previously available source.

In the early neolithic period, there were few resources that were not natural. Hunting-gathering and eventually agriculture were the dominant activities. As agriculture became more settled on particular lands, the concept of property rights to the more or less exclusive use of these lands became progressively more important. Over millennia an immense variety of forms of property rights has been employed and tested by different groups, tribes, nations, and civilizations. At the same time more and more natural resources were being brought into use to satisfy human wants and needs. Minerals that were never considered resources became eagerly sought out and fought over. Since mankind had developed a long history of dealing with property rights for land, most natural resources that could be definitively located on or under the earth's surface fell automatically under society's system of land property rights.

Fugitive resources, such as water, animals, and fish, eluded simple land prescriptions. Fish and animal wildlife (for food rather than scenic and psychological purposes) posed very difficult problems because they were in the commons for anyone to take. Elaborate hunting and fishing rights developed in different societies, but the unreliability of these rights over time and at the frontiers of neighboring societies has accounted for innumerable conflicts, vendettas, and wars. Often it was no one's interest to protect and sustain the animal populations unless a powerful ruler appropriated the wildlife to himself and brought it under his protection. For other fugitive resources, such as flowing rivers, it was natural to assign property rights as an extension of neighboring (riparian) land properties. Serious conflicts still resulted due to differences of interest between upstream and downstream users, or among coriparians on lakes.

Common pool resources fall between fully private and fugitive resources by the definition of user's rights. Examples of common pool resources are oil, natural gas, and groundwater, where many property owners on the surface may own the mineral rights beneath their own property but excessive abstraction would reduce the resource holdings of their neighbors. Note that oil and natural gas are essentially nonrenewable, while groundwater, if properly managed, is renewable.

Interestingly, clean air first became a common pool natural resource in thirteenth-century Europe. Soft coal was burned as a substitute for another natural resource, wood, which was becoming scarce due to deforestation caused by the demands of a growing population for increasingly more fuel and cropland. As coal continued to replace wood, the noxious levels of air pollution produced could no longer be tolerated. King Edward I of England is said to have introduced the death penalty in 1306 for burning soft coal within the boundaries of London (Hodges, 1973, p. 74). This was an interesting and innovative method of environmental regulation—command and control in its most severe form.

Table 4.1 gives examples of different natural resources showing how they fit into the renewable/nonrenewable continuum. This table

Table 4.1
Classification of Natural Resources

Alienable	Common Pool	Fugitive
Renewable (flow resources)		
Forests	Lake fisheries	Migratory wildlife
Closed pasture	Sedentary wildlife	River fisheries
	Groundwater	Water in streams
Settled agriculture	Open pasture	
	Nonfumigating air	Fumigating air
	Shifting agriculture	Solar radiation
		Tides
		Winds
Nonrenewable (stock resources)		
Private land	Public land	
Soil	Unique landscapes	Aeolian dust
Most minerals	Oil and gas	Sediment
Soil nutrients	"Old" groundwater	Nutrient leachate

also divides the natural resources into those that are easily privatized, those that are common pool resources, and those that are fugitive resources. The boundaries of these categories are sufficiently elastic that some resources may be categorized in more than one way depending on the size of the holding (if you own the entire lake then the fishery is not a common pool problem), on the definition of the resource itself (is it land, or soil, or soil nutrients?), and on the actual scale of the intervention (one windmill has little effect on the available wind power, but 1,000 make a difference because of aerodynamic interference between them, so that wind energy becomes more like a common pool resource than a fugitive resource). Water flowing in streams is generally considered a fugitive resource, whereas groundwater is usually considered a common pool resource. An airshed over a city may behave like a common pool resource if the air is stagnant and the pollutants build up (Los Angeles on a hot summer afternoon), or it could be a fugitive resource if the wind blows the pollutants (fumigation) from one region to another.

Even nonrenewable resources exhibit these same ambiguities. For example, soil can be owned outright as a nonrenewable resource, but the same soil if blown away by wind (aeolian dust) becomes a fugitive resource (and also the source of serious external damages to someone else's property). The same factors hold true for soil nutrients.

Theories Underlying Natural Resources Policies

Natural resources policy as a field of scientific inquiry may be defined as a study of actions of governments at different levels with reference to the intertemporal distribution of the use rates of resources. The study is concerned with the formulation, the objectives, the tools, and the effects of conservation policy, and with criteria that might be used to appraise performance, that is, the effects in relation to the objectives. (Ciriacy-Wantrup, 1963, p. 223)

It is important to note that although policy often uses the tools of economics, it is not necessarily primarily an economic endeavor. This point is often overlooked by modern commentators, who often assume the hardest parts of policy as given, namely formulation of goals and objectives, and concentrate their attention upon tools, effects, and criteria. The former deal with issues in political philosophy and sociology (chapter 7) and the latter with economics (chapter 6), law (later in this chapter), and engineering (chapter 5).

In order to explain natural resources policies it is first necessary to understand the basic assumptions made by policymakers. There are four main philosophical approaches to natural resources, each of which dictates how policy should be formulated. The first three approaches are broadly stated economic approaches and the fourth is based on "ethical natural science." Which theory is used, or believed, radically influences the policy tools chosen for resource management.

Theories based upon property rights.   Much of the discussion about natural resource use focuses on the question of property rights. Property has been called "the primary economic institution" and is of great importance to natural resource use, in its own right and because of several "derived" economic institutions (Ciriacy-Wantrup, 1963, p. 141) such as tenancy, credit, and taxation. Recent publications of the Pacific Institute for Public Policy Research have stressed a theory of efficient natural resource use based upon property rights (e.g., Anderson, 1983a; Stroup and Baden, 1983). Their theory appeals to the market and neoclassical theories in economics for its ultimate justification.

Theories based upon scarcity. The earliest theories about natural resources can be traced to two Englishmen with unlikely English names, David Ricardo and Thomas Malthus. Their theories at the end of the eighteenth and the beginning of the nineteenth centuries earned economics the sobriquet "the dismal science." The outlook of Ricardo and Malthus rests on three premises (Barnett and Morse, 1963):

1. Natural resources are economically scarce and become increasingly so with the passage of time because of population increase and the growth of the economy.
2. The scarcity of resources limits economic growth.
3. Ultimately economic growth is brought to a halt by natural resource scarcity or exhaustion.

The last premise is commonly referred to as Malthusianism. After almost two centuries of continued and accelerated economic growth despite Malthus and Ricardo, there are still periodic outbreaks of Malthusianism, or neo-Malthusianism, as exemplified by the Club of Rome's *Limits to Growth* (Meadows et al., 1972) or President Jimmy Carter's *Global 2000* (Barney, 1981). The scarcity theory of resource use does not seem to deal well with technology changes, adaptive human changes to use fresh combinations of resources, and the adaptive behavior of human populations with respect to their own reproduction.[1]

Theories based upon neoclassical economics.

There is no need for a new and grandiose theory of resources; what is needed is for society to apply its understanding of the theory of production and the behavior of the firm to a different situation. (Scott, 1955, p. 1)

In essence what Scott and others are saying is that neoclassical economic theory is required to formulate natural resources policy. As shown in chapter 6, however, there are a set of assumptions underlying the classical theory that may or may not be acceptable in many societies; the most troubling of these is the assumption that the prevailing distribution of income is socially acceptable.

Theories based upon ecology.

Principles derived from the natural sciences have been developed for the conservation of each resource—forests, ranges, soil, wildlife—one principle, that

of ecological balance, or nature's balance, for the several sources in combination. This latter principle states that under natural conditions the resources
of a region tend to be in balance and to remain relatively stable over long
periods of time. (Maass, 1968, p. 274)

The theories underlying the ecological approach to natural resources
policies rely heavily on scientific analyses of individual components
pieced together with a not altogether scientifically defensible idea of
nature's balance. All of this is overlaid with a strong Protestant ethic
of good stewardship for the common and future good. In the U.S.
body politic this is a powerful combination that allows for science
worship, moral indignation, and venting otherwise disembodied
frustrations. The combination of scientific elitism and positive government makes it hard to object to these theories. That these ideas
were initially propounded by rugged individualists like President
Theodore Roosevelt only made them more attractive. Later we will
see that if any theory currently has the upper hand it is most likely to
be this one.

The activist political philosophy of the early leaders of this approach, men like Gifford Pinchot who headed the U.S. Forest Service
for Theodore Roosevelt, has not had the backing of the president and
executive branch in recent years. Pinchot believed that it was the duty
of executive officers (including the president) to do everything for the
public good—as they interpret it—that is not actually prohibited by
law. This contrasts with a more sedate view of environmental matters
that has the executive doing merely what the law directs it to do.

The ecological theories provide a counterpoint to the economic-
based theories. In the debate about allowing effluent taxes, the eco-
logical-ethical approach stands in direct conflict with the economic
approach, which looks for economic efficiency through pricing and
taxation. Strangely enough, adherents of the ecological viewpoint do
subscribe to the economic theories of scarcity enunciated by Malthus
and Ricardo and often view themselves as trying to save the economy
from its ultimate collapse.

Development of Natural Resources Policy in the United States

Ciriacy-Wantrup suggested that the conservation movement in the
United States during the 1870s arose as a result of population growth,
rapid changes in technology and social institutions, and the realization

that the geographic limits of our country's expansion would soon be reached (Ciriacy-Wantrup, 1963, p. 14). The movement led to changes in the essentially laissez-faire natural resources policies that, at the general level, had been the rule since the first colonization. This does not mean that there had been no concern for rational management of specific natural resources—the earliest enactment of Massachusetts's colonial legislature, as has been mentioned, was a measure to regulate the fishery. But only in the 1870s did an understanding emerge of the broader ecological effects of various piecemeal policies and the gaps between them.

With regard to deforestation, America in the 1880s was similar to Brazil in the 1980s. The setting aside in 1891 of unreserved public timberlands as national forests for the protection of watersheds was the first important landmark in the development of the new policies. The flowering of federal natural resources policies came during the presidency of Theodore Roosevelt (1901–1909) when, aided and abetted by Gifford Pinchot, many federal specialist agencies were developed and activist executive agencies promoted conservation policies. Outside of the government the same trends were encouraged by prominent environmentalists like John Muir of California who were leading many local fights against exploitative development pressures.

From the early part of this century the development of natural resources policy and water policy were intertwined. Many of the forestry, fish, and wildlife actions by the federal government were carried out as part of, or in conjunction with, water activities. Even though influential agencies like the Forest Service, the Fish and Wildlife Service, the National Park Service, and the Soil Conservation Service were created, they never attained the political and budgetary importance of the traditional water agencies like the Army Corps of Engineers. Both water policies and the other natural resources policies received powerful impetus from the conservation movement and later from the environmental movement.

The ecological theory of natural resources policy stemmed from the late nineteenth-century conservation movement. This movement embodied some good ideas, performed major miracles in stopping the rape of America's landscape by unregulated industry, and above all attracted strong, intelligent leaders with a strong commitment to preserving the environment for future generations. Of course conservationists also had an eye to developing natural resources at some reasonable rate of exploitation ("wise use"). By the time of the New

Deal they wanted to "maximize the use of resources for the greatest good of the greatest number."

The two strands of preservation and wise use coexisted comfortably within the conservation movement for 90 years until about 1960, when the preservationists split off to constitute the environmental movement. So at one level it is correct to identify the birth of the environmental movement with the early 1960s, but the movement was born as a mature adult. The ten years from 1960 to 1970 were a period of continued population growth, rapid urbanization, and extremely rapid economic growth seemingly without end. The newly urbanized populations were in the most polluted and deteriorated parts of the country, and with their increased affluence the urban populations could start to demand improved living conditions. At the same time Congress was also becoming more urbanized, with the memberships of key natural resources committees reflecting less and less the old agricultural concerns with soil and water conservation and more the interests of their urban and suburban constituents in environment and recreation.

Imperatives That Influence Water Policy

There are elements inherent in all policy areas that must be attended to or the policies will not work. While these imperatives include the most basic laws of physics, chemistry, and economics, it is the interaction of these general principles with the specifics of a problem that create the important imperatives of that problem. "Water runs downhill" is a scientific imperative; "Water runs uphill toward money" is an economic imperative; and "Water quality goes downhill if upstream and downstream users do not cooperate" is a political imperative.

That imperatives can be ignored only at considerable risk to a favorable outcome is most easily seen for technological or scientific imperatives. For example, taking into account the entire system is a simple scientific imperative when one is dealing with complex dynamic systems. We have ignored it for many years in the water pollution business, for example when we chose to focus on the technology for controlling sources of water pollution rather than focusing on the quality of the ambient environment and asking what would be the best way of achieving a desired level of quality. We now find ourselves, many years and many billions of dollars later, still faced with major water pollution problems due to diffuse sources of pollution (non-point sources). Henry Maier, the former longtime mayor of Milwaukee, complained that after spending over $900 million on water pollution control since 1972 the city was still not able to use its waterfront for recreation and tourist development, because of pollution caused by non-point sources far outside the city's jurisdiction. He asked plaintively, "Why did nobody warn us of this 15 years ago?" It may take another $900 million expenditure on water pollution control before the city can develop its waterfront as it would like.

Mayor Maier and the city of Milwaukee have suffered a $900 million misunderstanding. There was an unreconciled conflict between the scientific and the political imperatives, and in this case the political imperatives, such as ease of enforcement and equitable application of the law, had won out over the scientific ones. In the United States we need to develop a framework that will highlight potential conflicts between differing imperatives and resolve them before legislation is passed, institutions are created, regulations written, and money spent, much of it fruitlessly for lack of early clear-headed analysis.

In the natural science of water, the hydrological cycle described in chapter 2 is a simple and powerful scientific imperative that, if obeyed, could give coherence to many policies for resource use and management. Taking account of the entire cycle should help us not to overlook the materials that flow into or out of our environment as a result of all our actions, including treating wastes. For example, water is usually considered a renewable resource that passes through the several phases of the hydrological cycle. One can start anywhere in the cycle and get back to the same point through all the other phases. It is also a fact that mankind has recently been able to make significant impacts on the workings of this cycle. By mining metallic ores, by chopping down forests, and by creating synthetic chemicals we have, over the last century, had major impacts on the pollutant load carried by the cycle. In addition, we have also made large diversions of the flowing rivers and pumped heavily the groundwater resources so that large changes have been made locally in the cycle. Considering the cycle as a whole should help clarify the relative magnitude of contributions to the environmental problems. It would also be harder to fall into the errors implicit in regulating one effluent by converting it to a less regulated pollutant in another medium. (For example, incinerating or land-filling sewage sludge simply moves the problem from water to the air or the land.)

Technology itself has strong imperatives. For example, depending upon the least-cost technology available, technologists may recommend that a pollution problem be dealt with: (1) at the point of generation either by "end-of-pipe" treatment or by making a process change to eliminate the production of the polluting waste; (2) in the environment itself, by treating either the ambient surface or groundwater; or (3) by treating the water when it is withdrawn for use. This choice has many consequences for the administration and implementation of control systems. There are also many choices to be made as to whether to use capital-intensive methods or operation- and management-intensive methods. Depending on the choice made, many financial, economic, ecological (or perhaps environmental), political, and institutional adjustments are necessitated or ruled out.

One of the lessons that we learn from economics is that if something is worth doing, it's not worth doing well. This seeming misquotation of an old adage highlights the economic concept of marginal value. In other words, because some intervention is technically and economically feasible does not mean that it should be deployed to its fullest extent. There typically is some level of deployment, far short of perfection, that balances the benefits gained from deployment with the economic costs. Economic imperatives deal precisely with balancing technological costs with the benefits received. The demand for water, or for water of high quality, is not an absolute end in its own right but is a function of the incomes of individuals, the weight they give to water in comparison with their demand for other goods and services in the economy, and the actual costs of each one of the available goods. The producers and consumers of water try to maximize the net benefits associated with using water and preserving its quality.

For individuals and local groups, financial imperatives may be difficult to distinguish from economic imperatives. Unfortunately, there is often a direct conflict between economic and financial imperatives. Since the costs and the prices in the economy are often distorted by taxation, subsidy, resource scarcity, physical and economic externalities, etc., the individual producer or consumer will not be able to act on the correct economic signals even if he or she wishes to. Following the financial prices could lead to quite different behavior on the part of the producers and the consumers than would be "correct" under economic pricing. Consumers purchase commodities on the basis of the prices they see, not of the unsubsidized price, hence socially preferred solutions may not always arise on the basis of market forces. In the presence of financial and economic disparities and pervasive economic and physical externalities, it is usually not possible to achieve economic solutions of broad and long-range efficiency without some form of government intervention and/or regulation, which introduce a set of political imperatives.

Water as a Renewable and a Nonrenewable Resource

Earlier in the chapter the possibility was raised that it may not always be possible to decide a priori whether a natural resource is renewable or not; water is very much such a resource. The traditional view of water is that it is renewable. In other words, water is not used up by human and other users in the environment—it always becomes available again at some point in the not too distant future. At the global level very little water leaves the planet Earth, insignificant amounts are involved in chemical dissociation into hydrogen and oxygen, and only very small amounts are sequestered in sinks that do not eventually return to the hydrological cycle. Water is seen as the most important of nature's grand cycles. (Other important ones are the carbon, nitrogen, sulphur, and phosphorus cycles [Smil, 1985].) The annual water cycle is the most important in part because it is the medium for most of the transport and diffusion of the other important chemicals throughout their cycles.

Groundwater often appears to be nonrenewable. This is because it is difficult to assess when the resource is being overstressed, since it is out of sight and the mechanisms of recharge and participation in the hydrological cycle are complex. Groundwater exploitation by many private parties acting at the same time without coordination can lead to overdrafting until the groundwater is apparently used up. Then use declines and the natural recharge will eventually refill the aquifers. This is a nice example of a natural homeostatic mechanism; it can be seen in the Ogallala aquifer, which runs northward from Texas to Nebraska. For many years, in places like the Texas panhandle, this aquifer has been seriously drawn down due to the pumping of irrigation water for crops such as cotton (which grows in abundance under rain-fed conditions across the Mississippi to the east). Local water table declines of as much as three to five feet per year have been observed, with total declines this century amounting to as much as 150 feet. This has made irrigation increasingly expensive and many farmers are returning to dry farming. When the pumping stops, the natural recharge will ultimately replenish this aquifer—to be pumped down in some future era by farmers convinced that the water below their land is a purely private resource. In the meantime, the Texas congressional delegation could not convince the Congress as a whole that the behavior of the Texas farmers was something that should be supported by diverting water from the Missouri and Arkansas rivers to compensate for the groundwater exhaustion. Indeed, even the Texas taxpayers soundly defeated a proposal to have *them* pay the $5 billion to divert the river water. Nature and the Texas taxpayers were teaching the farmers something important about ecology: *just because something is possible does not mean that it has to be done.*

But groundwater and surface water can become so badly polluted that for all practical purposes they are unusable. This is particularly true of groundwater and is the basis for many of the current initiatives to control groundwater pollution. Chapter 6, however, demonstrates that it is simply an economic decision whether cleaning the water to a given level of purity is worth the cost. There is nothing that we add to water, either intentionally or inadvertently, that cannot be removed at some cost. The assessment of a given source of water as renewable or nonrenewable, therefore, depends upon the use to which it will be put and the cost of replacing or cleaning it. Seawater is usually considered an uneconomical source of fresh water, but in the Middle East water is precious and energy is cheap. For example, in Saudi Arabia, where oil and natural gas are abundant, 26 desalination plants operate to produce 500 million gallons of water per day for a variety of purposes.

Water as a Public or Private Good

Most people in everyday use do not regard water as an end in itself. It is a commodity, consumed directly or used as an input to other processes. The problem is that most people understand water this way at an individual level but treat water differently when they gather collectively to make decisions about the future use of water. This may be due to the common social misperception of water as a common pool resource, which belongs to everyone at once with a right of access for all. As a result, it is customary to treat water as a free good when it is in fact anything but free. People come to expect that it is their right to take (and to waste) as much water as they want.

There is clearly a paradox here, since it is obvious that for many uses (for example irrigation and municipal supply) water has all the properties of an exclusive economic good, that is to say, just the opposite of a public good. Adam Smith was the first to make a definition of a pure public good. The example he used was that of lighthouses and other navigational aids—which were among the first federal investments made by the fledgling U.S. Congress in 1789. Once a public good is provided it is not possible to exclude anyone from using it who wants to take advantage of it, and the consumption of the good by any one consumer does not impede consumption by any other potential consumer. At the other extreme a pure private good, such as food purchased from a supermarket, can be and usually is the exclusive property of an individual who can exclude others from consuming it, and his or her consumption of the food absolutely prevents anyone else from consuming it. Clearly, water like many other goods falls somewhere in between these two extremes. It has aspects of a pure public good when it is left in a scenic river, but even then too many people using the scenic river will destroy some of its value for other participants. It has aspects of an exclusive economic good when it is evaporated by farmers irrigating their crops for profit in a market setting.

It is the in-between cases, however, that tend to predominate in the use of water and that confuse the consumers. If I use water and return it to a river for someone else to use downstream, is it then a "public good"? It depends upon whether I return the same quantity and quality of water that I abstracted from the river in the first place. If so, the water has aspects of being a public good. But if my access to the water source required my owning the riparian land on the river, then the water certainly was not a public good. The question of who has access becomes very important and hinges upon the different doctrines governing water rights discussed in some detail below. An important case of this ambivalence is that of public provision of water transport facilities. Chapter 6 documents an attempt to make the users pay for the use of such facilities.

Doctrines of Water Rights

Maloney and Yandle articulated a "Theory of Alienability" for water based upon a search for wealth created by scarcity (Maloney and Yandle, 1983). Natural resources are "alienated" from the environment according to four propositions:

1. As a resource becomes more valuable, more effort will be devoted to property rights definition and protection.
2. Contract enforcement costs are a function of the structure of the contract.
3. Increasing the generality of property rights, i.e., increasing the flexibility, discretion, or expansion of the rights transferability by the holder or contracting parties, causes the cost of definition and enforcement to rise.
4. The more alienability included in property rights, the more valuable they become, but at a decreasing rate.

The important terms in these propositions are definition, transferability, and protection of the rights. Maloney and Yandle claim that the above propositions explain the past and serve as the correct underpinnings for policy prescriptions. They believe that over time one can observe a shift, in response to their propositions, along a sequence of four basic property institutions:

Institution	Example
Common access without scarcity	No rationing, free access. Early land settlement in the U.S. West.
Common property	Rationing by excluding non–group members. Tribal hunting grounds.
Public property	Exclusionary rules, user privileges, and enforcement. National parks.
Private property	Fully tradable, fee-simple rights. Private residential real estate.

Roughly speaking, as resources become scarce, common access ceases to be attractive and the resource then tends to be held as a common property of a group that excludes some users, or potential users. As demand increases for use of the resource, social pressure grows to have it become public property, to which the state regulates access, or be defined as private property, to which individual owners regulate access. This transition explains why Hardin's "tragedy of the commons"—in medieval Europe everyone was entitled to graze animals on the common lands and there was no incentive to prevent overgrazing—very rarely leads to the predicted overexploitation of the commons: long before that stage is reached, individuals, groups, and society act to change the nature of access, by force if necessary.

Stroup and Baden use these concepts as a framework for the discussion of the "best" approach to property rights for efficient allocation of society's resources (Stroup and Baden, 1983). Stroup and Baden are a part of a large movement that recommends placing all natural resources under the rule of individual ownership. They go even further, stating that "a theory of property rights can become a theory of the state" (p. 7), and that "private ownership of property rights alone is sufficient to secure efficient resource use" (p. 17). Given the modern renaissance of states based upon theocracy, racial identity, and other emotional claims under the banner of nationalism, the first assertion seems generously exaggerated. And awareness of the problems of public goods, externalities, and the sharing of common pools associated with many natural resources tempts one to respond "necessary but not sufficient" to the second assertion.

Markets are based on a system of property rights to scarce goods and the right to exclude other users of the resource. A private market only exists, however, if the property rights can be transferred. Hence, private transferability of property rights is the distinguishing feature of markets, as opposed to other methods of allocating scarce resources.

In the United States the states are "sovereign" over water, and hence, with a few exceptions noted below, any discussion about property rights has to deal with state law. At any level, however, the definition of property rights in water is very difficult in the face of pervasive externalities, and a large body of law has developed around water. Getches (1990) provides an excellent overview of both surface water and groundwater rights applied in the 48 contiguous states, and is the source of the discussion below.

Surface water rights.   Three major surface water rights doctrines hold sway in the United States: riparian rights, prior appropriation, and hybrid systems. Riparian rights appear to be largely derived from English common law and the French civil code. In the relevant periods, both of these countries had little irrigation or other large water-consuming activities; it could therefore be assumed that most of the water withdrawn would find its way back to the stream. Currently, 29 states utilize the riparian rights system, whereby a landowner receives limited rights of water *use* (not possession) for water bordering on his or her property.

The prior appropriation doctrine, followed by 19 states—strictly in 9—is a uniquely American conception arising out of the Hobbesian anarchy of mining camps in the American West during the 1840s. The miners mainly squatted on federal land and needed large quantities of water to use hydraulic mining techniques on their claims, which were considerable distances from the rivers. In the absence of organized law enforcement, they devised their own rough and ready rule of "first in time, first in right." The prior appropriation doctrine continues to function on the basis that the first entity to use a water source claims sole rights to its use.

The third major doctrine, the hybrid system, is a combination of the other two doctrines. The hybrid system is established in 10 states, and was formulated primarily to introduce the benefits of prior appropriation to historically riparian states. By combining the best parts of the riparian rights doctrine with the prior appropriation doctrine, the efficiency of the overall system improved.

These three main doctrines are moderated by other reservations of water by the federal government, and by the more recently developed public trust doctrine now being applied by some states. This is a flexible doctrine stipulating that public natural resources, of which the government is a trustee, must be maintained for public use unless alternative restrictive or private uses of the resources benefit society.

Groundwater rights.   Groundwater rights are more complex than surface water rights because of the common pool nature of most groundwater sources. Moreover, it is only within the last 40 years that the possibility of serious groundwater depletion has arisen, because of the advances in the technology of well drilling and pumping that played a large part in the post–World War II expansion of irrigated agriculture. Five groundwater doctrines are currently in use across the United States: absolute ownership, prior appropriation, groundwater as a public resource, reasonable use, and correlative rights.

The absolute ownership doctrine is applied in 11 states; it simply grants possession of water found under private property to the landowner. The prior appropriation doctrine is the groundwater counterpart to the surface water doctrine of the same name—the first person to use a specific groundwater source acquires defendable rights to its usage over subsequent users. However, most states treat groundwater as a public resource, meaning that it cannot be privatized. Even states that do not consider groundwater a public resource rarely allow absolute ownership. Instead, they turn to the reasonable use doctrine, which allows a landowner to withdraw water for reasonable beneficial uses on the overlying land without liability for harm to adjoining landowners. This has become known as the "American Rule" (Getches, 1984, p. 238). Yet another doctrine regulating private use of groundwater stems from correlative rights. Correlative rights takes the reasonable use doctrine one step further by allocating each user's reasonable share of available groundwater according to each individual's land acreage over the groundwater source.

In the West the major unresolved water policy issue is that of federal and Indian reserved rights. The Indian rights stem from treaties between the federal government and the defeated Indian tribes. The Supreme Court ruled in 1908 that these claims were to be based on the timing of the establishment of the reservations, which in most cases predate the arrival of white settlers in the West. Since these rights were often not used and not claimed by the Indians and were subsequently used by later settlers, many water rights in the West are not certain. Based upon these arguments, Indian water rights in Arizona alone more than exceed the entire flow of the Colorado River. Getches explains the issues as follows: "The reserved rights doctrine was created to assure that Indian lands and public lands set aside by the government for a particular purpose would have adequate water. The doctrine recognizes rights to a quantity of water sufficient to fulfill the purposes of the reservation. This doctrine is known as the Winters Doctrine after a Supreme Court decision in 1908" (Getches, 1984, p. 291).

The property rights doctrines also extend to cover water quality and its safety as a drinking water source. The concept of a physical quantity of water as property is easier to understand than the ideas of "quality" and "safety" as property rights. Table 4.2 shows how property and the property rights were created (or taken) by the 1972 Clean Water Act. At least three different types of property were cre-

Table 4.2
Property Rights Established by the 1972 Clean Water Act

Defined Property	Implied Owner
Baseline of quality described by effluent limitations	Federal government
BPT level of pollution with variances by 1977; BAT with variances by 1983	Existing polluters
BAT level of pollution on permissible waters	New polluters

Source: Based on Maloney and Yandle (1983), p. 304.

ated by this legislation, enjoyed by three different classes of owners. The first property is "a baseline of water quality described by the effluent guidelines," which was given in the name of all of the citizens to the federal government. Existing polluters received the right to pollute up to the so-called best practicable technology currently available (BPT) level until 1977, meaning that they had to use the best practicable technology to treat their wastes, and then at the best available control technology (BAT) level thereafter, meaning treatment must use the best available technology economically achievable (generally a higher level of waste removal that is more expensive than BPT). The third beneficiaries were the new polluters, who are allowed to pollute up to the BPT level (until 1983 and BAT thereafter)[2] only on the "effluent-limited," not the "quality-limited" streams (the latter are waters for which the amounts of added pollutants are limited by ambient water quality standards). In quality-limited reaches of the river, new polluters would be required to meet much higher new source effluent limitations. As can be seen, regulations produce winners and losers with asymmetrical property rights; property rights were created for the general public who had none before, the property rights of the existing polluters were reduced to the new levels, and newcomers' property rights to pollute were set at an even lower level than those of the existing polluters. This property creation effect occurs whenever government regulations are introduced into any previously unregulated area of concern.

Constitutional Basis for Federal Intervention in Water

A purist studying the Constitution would note that there is not one reference to water or waterways in it. So how can we say that our water resources policies are based upon the Constitution? (Schad, 1988, p. 3)

The different state-level legal doctrines pertaining to water are sufficiently vague in places to ensure continuing litigation on many water issues. Each of the doctrines is implemented by state laws that are often similar in different states but not similar enough to make them of global application. Overlaid on these doctrines of rights to certain volumes of water are also explicit and implicit water quality rights. From a federal perspective, Schad claims that there are five provisions in the Constitution that provide the basis for water policy: the power to regulate commerce, the power to take action with regard to the territory of the United States, the power to levy taxes for the general welfare, the power to declare war, and the power to make treaties with foreign governments. The most important powers, and the ones most often cited, are those that derive from the commerce clause of the Constitution:

Section 8. The Congress shall have Power . . . To regulate Commerce with foreign Nations, and among the several States, and with the Indian Tribes. . . .

Even though the applications are probably far from what was originally intended by the framers of the Constitution, this clause has become the constitutional basis of most federal water legislation. Originally intended to regulate interstate commerce, the clause has been successively interpreted by Congress and the Supreme Court to regulate activities wholly carried out within one state but which have effects on interstate commerce. Firestone and Reed (1990) give examples of the tenuous paths by which disposal of rubbish at a local dump site may set off a chain of events such that after a rainstorm runoff from the dump pollutes a local lake, polluted lake waters have fewer sport fish, fewer sport fish means fewer fishermen who buy less gasoline and fishing tackle, both of which are articles marketed in interstate commerce—thus the local rubbish disposal affects interstate commerce and can hence be regulated by the federal Congress if it deems it necessary.

More direct justifications are used for many regulations on surface water use and its pollution. Inland navigation was the major form of transport and hence of interstate commerce in the early days of the country, when the Constitution was being written. Therefore the federal government was deemed empowered to regulate all navigable waterways. The word "commerce" was construed by the Supreme Court to cover the word "navigation." This was first used to regulate the navigation itself, then clearing of the rivers; later it was the justification for flood control, landfilling, and subsequently water quality measures. Over time, the definition of what are "navigable waters" has also become broader and broader until it now encompasses "the waters of the United States." Firestone and Reed cite one case, *United States v. Holland* (373 F. Supp. 665 [1974]) (Firestone and Reed, 1983, p. 76), in which the defendants admitted discharging pollutants into man-made mosquito control canals in wetlands without a permit, which the government claimed was required under the Clean Water Act. The defendants argued that the receiving waters were not within the federal jurisdiction. The court ruled that what was meant by "navigable waters" was not necessarily what it would mean in the technical sense but rather what Congress had intended it to mean.

The Safe Drinking Water Act had its dim antecedents in the formation of the U.S. Public Health Service in 1912 and the Public Health Service's subsequent regulation of the drinking water on boats and trains involved in interstate commerce. The Clean Water Act (particularly Section 404) similarly has its antecedents in the Refuse Act of 1899, which was originally intended to restrict dumping of spoil materials in navigable waterways (hence interfering with interstate commerce).

Superimposed upon the constitutional arguments for a federal interest are the congressional rationales for federal government activity, which have grown over time. These have been identified by Mugler as public/private interrelationships, federal responsibilities to society, and intergovernmental relations (Mugler, 1982, p. 2).

Mugler argues that private barriers, such as uncertainty and risk, and private disincentives call for federal intervention in the private sector to improve environmental decision making on water issues. This constitutes a public/private interrelationship. Furthermore, the government has responsibilities to society and therefore must have an interest in these issues; for example, many water-based issues transcend boundaries, thus requiring assistance beyond the local level. Also, severe emergency situations often require aid that can only be provided by the federal government (e.g., federal floodplain zoning). However, the largest subset of the congressional arguments for a federal interest in water fall under the rubric of intergovernmental relations. This is a clue to the current levels of unhappiness with federal water policies and their implementation by the nation's governors.

Several factors are involved in federal interest through intergovernmental relations: governmental barriers and disincentives, the variable quality of state and local programs, precedence in federal programs, and the need for cost sharing by the federal government.

Current Federal Water Legislation

Given the widespread federal involvement in water over the entire history of the United States, it is no surprise that the current federal legislation governing water use is an extremely large, fragmented, and complicated body of literature. Since the Constitution was adopted in 1788 there has been a steady accretion of specific pieces of federal legislation—in recent times more like a torrent than a glacier. Over 50 federal statutes deal with water and its environmental impacts. It is relatively easy to identify the enactments directly relevant to the three areas of water resources, municipal and domestic water supply, and water quality and environment, because the basic legislation addresses these areas separately. It is much more difficult to identify other laws that deal only peripherally with water's impacts or those that inadvertently place restrictions on water use in the quest to regulate some quite different social or economic phenomena.

While in earlier times fragmentation was understandable, and even useful in its effects, it is no longer the appropriate way to deal with many of the water issues that are now correctly viewed as multimedia (air, land, and water) or otherwise cross-cutting issues. For example, the most difficult problem in the $6 billion cleanup program for Boston Harbor is to dispose of the sludge produced by the wastewater treatment plant. Depending upon the technology suggested this becomes either a solid waste disposal problem or a problem of air pollution—neither of which is specifically covered in the relevant water legislation. Nationwide, many groundwater contamination problems arose because different agencies operating under different legislation and different guidelines dealt separately with wastewater and with solid waste disposal but no authority (nor any legislation) was concerned specifically with groundwater protection—hence the problem went unnoticed for many years.

Water resources.   In December 1986 the Water Resources Development Act became law. This was the first omnibus water bill to be enacted in 10 years and the first serious water bill in 17 years (Cortner and Auburg, 1988). In part it was a continuation of the pork barrel approach of the earlier omnibus water bills, but it also represented a major departure from the previous water resource legislation. The policy provisions of the act, as summarized in chapter 3, required an equal financial contribution by a nonfederal sponsor for civil works projects, cost-shared planning, an increased nonfederal cost share for project construction, and up-front nonfederal financing during actual construction. Environmentalists were encouraged by the act's beneficiary-pays principle, which promised to limit the most unsatisfactory water projects (Foster and Rogers, 1988, p. 32).

The Reclamation Projects Authorization and Adjustment Act of 1992 represents the major shift that has been taking place in water resources development in the United States. This act lays out clearly a mandate for ecosystem protection via *changing* the operation of previously developed federal water projects. While the emphasis in the act is on California and the West, the basic principles seem to be now well established and should be applicable in future legislative action to all federal water activities.

Municipal and domestic water supplies.   The growing concern for contamination led to the Safe Drinking Water Act, enacted in 1974 and amended in 1986. Under the amendments the Environmental Protection Agency is now required to promulgate national primary drinking water standards and to specify the maximum levels permitted for 83 contaminants potentially injurious to public health. The states have the primary responsibility for taking remedial action. The passage of the water conservation parts of the Energy Policy Act of 1992 should have a major impact over the next 20 years on the demand for municipal and domestic water supplies. Also, due to the reduction in flow it will have a major impact on wastewater generation and consequently upon water quality.

Water quality and environment.   The Clean Water Act amendments of 1987 represents one of the major pieces of federal legislation in the whole water area. Starting as the Water Pollution Control Act of 1948, and modified in 1956, 1965, 1972, 1977, 1981, and 1987, the federal water pollution control program has grown from a modest presence to ensure consistency in state ambient water quality standards into a dominant actor in defining and directing water pollution abatement efforts across the country. Over this period, the emphasis has shifted away from water quality-based standards (defined in terms of ambient water quality) to those that are more technologically driven (setting effluent limits by treatment process). The federal government, through the EPA, now sets national water quality goals, requires the states to prescribe standards and establish priorities, and furnishes financial assistance (originally grants but now loans—see the discussion of this issue in chapter 3) to help achieve the targeted water quality improvements.

Other legislation. A comprehensive approach to environmental quality is represented by the National Environmental Policy Act of 1969. Under this act (and the various state equivalents), a formal environmental impact statement is required for any major action significantly affecting the quality of the environment. This procedural requirement is enforceable by citizen suit and subject to injunctive relief imposed by the courts. At the federal level, oversight is provided by a presidentially appointed Council on Environmental Quality.

The growing national concern about the contamination of water and the environment by toxic chemicals has prompted a number of other specific legislative measures. Provisions of the Toxic Substance Control Act, enacted in 1976, govern not only the use and disposal of toxic chemicals but even their manufacture. The Federal Insecticide, Fungicide, and Rodenticide Act of 1974, and its subsequent amendments, call for the registration of pesticides and the regulation of their use. Under it the states are given the primary implementing responsibility. The Resource Conservation and Recovery Act of 1976, and subsequent amendments, authorize the tracking of hazardous wastes from manufacture to ultimate disposal. Its provisions, and those of counterpart state legislative acts, govern all aspects of generation, transportation, storage, and disposal of hazardous wastes. The Comprehensive Environmental Response, Compensation, and Liability Act of 1980, setting up what is known as the Superfund, was a specific response to the Love Canal and other waste site revelations at that time. This act (and its subsequent amendments) is designed to facilitate the cleanup of individual hazardous waste sites. Large sums of money, derived from special taxes on waste producers, fines and settlements, and cleanup cost recovery from pollution offenders, have become available at both federal and state levels, often taking the form of revolving funds.

Several other laws dealing with aspects of water resources also contain specific action forcing standards or other requirements to protect the environment.[3]

Cross compliance with other federal legislation. An area of growing interest is that of advancing federal water and environmental quality goals through federal regulation in other areas of concern. For example, the Food Security Act of 1985 has provisions against "sod-busting" (bringing marginal land into production) and "swamp-busting" (draining wetlands), along with restrictions on erodible soils that, if strictly enforced, could have major impacts on non-point source water pollution problems and also on many wildlife issues. More conventional cross compliance involves proper coordination of the requirements and implementation of the Clean Air Act of 1977, as revised in 1990, and the Resource Conservation and Recovery Act so that problems of waste sludge obtained from sewage treatment do not lead to major land or air contaminants. Also more creative uses of the legislation in tandem could allow states and localities to take advantage of specific technological options that would better fit the local conditions.[4]

Federal Institutions with a Role in Water Policy

This brings us to the question of how federal water policy is handled at the present time—in short, whose responsibility it is. Thirty-five federal agencies in ten cabinet departments, eleven independent federal agencies, plus four agencies in the Executive Office of the President, five river basin commissions, the federal courts, and two bureaus currently carry some responsibility for water programs and projects. Appendix 4 lists these different federal entities, which often operate under individual legislative acts, rules, regulations, and executive orders. At least 25 separate water programs, and some 76 separate congressional appropriations accounts, have been identified.

The nation's primary water policymaking body, the Congress, is equally fragmented. By the 102nd Congress there were 14 House committees with 102 subcommittees, plus 13 Senate committees with 82 subcommittees, exercising responsibility over some aspect of water resources development and management (appendix 5). It is no surprise that the legislative enactments over the years have exhibited overlap, duplication, and even inconsistency.

These lists of course do not mean that each of the federal entities is equally important or that each congressional committee and sub-committee deliberates on each piece of water legislation. If this were the case, nothing would ever get done. Six of the federal agencies listed in appendix 4 account for the bulk of the employment for water and water-related activities. These are the U.S. Army Corps of En-gineers (39,000 employees), the Tennessee Valley Authority (33,000), the Bureau of Reclamation (7,300), the Geological Survey's Water Resources Division (4,300), the EPA (2,900), and the Soil Conserva-tion Service's Watershed Projects Division (2,000). While the remain-ing agencies account for few staff persons they can be very important in some of the regulatory or implementation aspects of water policy. Four agencies account for the bulk of the federal water expenditures.

Technology, Technical Fixes, Technological Imperatives

Much of the literature and discussion about water is highly technical and is frequently colored by the economic and social biases of the technical experts. Others involved in water policy and water management may be hard pressed to distinguish between imperatives strictly demanded by the science and biases introduced by the technicians themselves. It is, therefore, important to understand the basic scientific and technical concepts of the field.

One concept that has been heavily criticized by environmentalists is the "technical fix." In essence, a technical fix is the response to a crisis brought about by failure or misuse of technology, which is then "fixed" by some other application of the same, or different, technology. To an engineer this is a manifestation of how reliable and flexible his or her technology is, backed by an immensely optimistic belief that an answer can be found to all problems. To many nonengineers it has the feel of trying to patch up something with something that has just failed. At the microlevel (individual pieces of equipment) a technical fix is obviously the best response, but at the system level it may lead to trouble. If your car breaks down on the freeway the obvious thing to do is to have it repaired; but if the freeway system breaks down because of overloading, the solution may not be to build more and bigger highways. Rather, new concepts and technologies of "people moving" may be required. Environmentalists point to nuclear power plants as examples of technical fixes run riot—every system has another system to check it because errors were found in the original system, and so forth in layers upon layers of fixes.

Federal and other agencies responsible for water supply and management have historically relied upon technical fixes. In response to increased demands for water and water-based services the agencies proposed bigger and better facilities—more dams, larger canals, higher levees, deeper channels, more and bigger locks on the rivers, and more and more complicated water treatment plants—all heavily subsidized by federal and local taxes. The environmental viewpoint is that systemic fixes would be a better response, such as source reduction of wastes or demand management by rationing or pricing. These would reduce the need for increasingly complex and expensive technical solutions.

The conflict between these two approaches may never be fully resolved because there are good arguments on both sides. The resolution is ultimately a political one, whereby society decides that it is willing, or unwilling, to pay for the more painful social adjustment of a system fix over the easy technical fix.

Demands for water have generated some of the largest construction projects ever undertaken. Colossal projects such as the Aswan High Dam, the Hoover Dam, the Central Arizona Project, and the levees along the Mississippi are just a few of the mega-projects specifically designed to cause major changes in the underlying ecosystems. Considering the objective, it is not surprising that there are also often unintended negative impacts on aquatic systems. In addition, inadequate allowance is often made for unintended (and adverse) impacts on the larger ecosystems and human populations.

For example, irrigation projects have caused the spread of schistosomiasis, a debilitating infection by a parasitic worm that affects an estimated 200 million people worldwide, and malaria, which affects even larger numbers. Even activities not classified as water development projects may detrimentally alter aquatic ecosystems. For instance, agricultural and construction activities lead to soil and nutrient losses, and the nutrients are transferred to lakes and streams. Overpumping on some coastal aquifers in Israel, India, and the United States has led to permanent salination of the aquifer due to seawater intrusion. Failure to allow, and perhaps compensate, for such impacts can be costly indeed.

Human use and abuse of water supplies can convert once-valued streams and lakes into unproductive and hazardous bodies of water. If a society is to benefit in the long run from its resources and natural environment, it is essential to understand how human activities affect the aquatic ecosystem.

In the United States it was only recently that a broader understanding of the consequences of potential water projects was forced into the consciousness of planners by the passage, and the subsequent strict implementation, of the National Environmental Protection Act of 1970. The act raised the awareness of all the major water agencies, and of the general public, that it is not possible to construct large projects without having negative impacts on some aspect of the environment. This new awareness is one reason that the current time is of great interest to water policy analysts; for the first time, the past is no longer a good guide to the future because past approaches are now all but ruled out for many water uses. This point is underscored by the EPA's 1989 decision not to allow the City of Denver to proceed with the Two Forks Reservoir to supply water to Denver, even after a $40 million environmental impact study and approval by the Corps of Engineers.

Some of the most difficult technical and scientific choices facing water policymakers are in the area of setting quality standards for drinking water and wastewater disposal. Drinking water in particular takes decision makers into poorly understood areas of toxicology and epidemiology. It also takes them into the heart of the symbolic concerns for "purity" that make water different from most other natural resources. The only simple technical fix—that of completely eliminating all potential human carcinogens and other toxic chemicals from drinking water—is neither medically necessary nor economically feasible. So the decision makers are left to equivocate and quibble about concentration levels of pollutants in parts per billion (1 part per billion is equivalent to one teaspoon of material dissolved in an Olympic-sized swimming pool).

Basic Technologies for Water Control and Water and Wastewater Treatment

The basic technologies for water control have been known since antiquity; modern science has added little to what the ancient Egyptians, Mesopotamians, and Romans knew about providing water either for irrigated agriculture or for domestic and municipal uses. Inundation irrigation practiced by the priest-engineers of Egypt was established by 5000 B.C., and clay tablets from 2500 B.C. refer to large diversion canals on the Tigris and Euphrates, and at Mohenjo-Daro and Harappa on the Indus. The municipal water supply of imperial Rome, which provided protected water supplies from faraway sources by means of aqueducts and canals, incorporated such advanced technology as huge inverted siphons to cross ravines hundreds of feet deep; it was, and still is, a wonder of the world. It also included waterborne waste disposal systems not unlike many still being designed in many parts of the world, and the main sewer (the Cloaca Maxima) is still incorporated into the present Roman wastewater system. Large masonry dams were introduced about the time of the Romans and many survived for many centuries (the oldest surviving masonry dam was built c. 100 A.D. in Prosperpina, Spain).

The old technologies typically rely on gravity to deliver the water to the users and carry the wastes away to their ultimate disposal. However, by the fourteenth century the Europeans had developed many human-, animal-, wind-, and water-powered pumping systems specifically aimed at overcoming the limits of gravity in the movement of water and waterborne wastes (Agricola, 1950). The motivation for the development of steam engines in the late eighteenth century was the drainage of mine water. Pumping of municipal water in Philadelphia in 1801 was one of the first public uses of steam engines in the United States.

The initial task of waterworks was merely to move water from where it was to the point of use. Later it was realized that by building storages (basins, reservoirs, detention ponds) the water availability from seasonally fluctuating rivers could be stabilized. This led to the construction of storage reservoirs in upstream reaches with earthen and later masonry dams. The modern science of hydraulics, largely developed in eighteenth-century France, added the ability to measure flows and the forces involved in the diversion and movement of water. These developments allowed for the construction of larger and more reliable structures for water diversion. In most water infrastructure design we have moved little beyond the eighteenth-century French scientists and engineers. We have incorporated new materials and modern mechanized construction techniques, but little else from twentieth-century science.

In municipal drinking water treatment, little has been added since widespread application of sand filtration (1860s) and chlorination (1910s). Wastewater treatment plants are little changed since trickling filters (1900s) and activated sludge units (1920s) were introduced. A water engineer of 70 years ago would find few surprises if he walked into many water or wastewater treatment plants today. This points to one of the major problems in water quality, namely that we are expecting our treatment technologies to produce levels of purity that were undreamed of when the treatment processes were first developed. Even 15 years ago measurements in the parts per billion were rare in the water chemistry field. Now we routinely expect these minute levels of measurement *and* expect our treatment plants to respond with appropriate levels of refinement. It is an amazing tribute to the engineers of bygone eras that their treatment technologies have performed so remarkably well under conditions that they could never have imagined.

The conventional approach to water supply, management, and disposal for all uses is composed of four steps: first, a water source must be located and protected; second, the water must be transported from the source to the user, and maybe storage provided at both the source end and the user end of the transportation system; third, the water must be distributed to the users; and fourth, the wastewater must be disposed of in some manner, possibly by recycling. This does not mean that for all purposes each of these steps has to be taken; for example, many industrial users are self-supplied by local sources and dispose of their liquid wastes by various on-site methods so that none of the off-site storage or transportation and disposal is required. A generic outline of a typical water supply/user/disposal system is shown in figure 5.1.

Original source.   The original source can be from a river, lake, or the ground. The identification of the potential of the source to supply the needed water is the realm of hydrologists (or geohydrologists in the case of groundwater). Careful scientific measurements over long periods of time are necessary to establish confidence that the source will be able to supply the need for some specified level of reliability. Depending on the location of the source with respect to climatic regions, and on whether it is groundwater or surface water, the amount that can be "safely" developed varies widely. For example, in the moist eastern part of the United States the amount of streamflow that can be safely abstracted 95 percent of the time (a typical engineering rule-of-thumb reliability level) is typically close to 75 percent of the mean flow, whereas in the arid West less than 15 percent of the mean flow could be available at the same level of reliability. For groundwater, because of its slower reaction time and the possibility of drawing the water table down in one year to be replenished in the next, it is possible to plan for levels of reliability closer to the mean recharge rate.

Once the location and amount of water available have been established, technology must be employed to abstract the water from

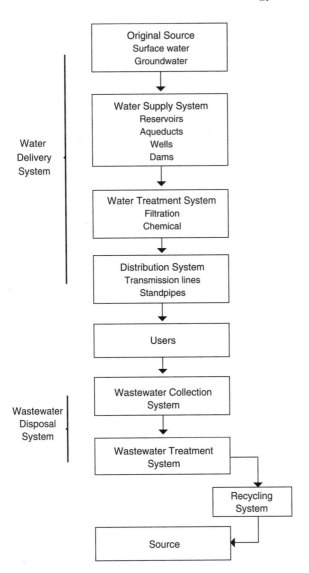

Figure 5.1
Water systems technology.

the source. For surface sources the layman typically thinks of a large dam as the means of diverting water—not an unreasonable supposition given that in the United States alone there are more than 60,000 dams, of which more than 1,000 could be called large (over 50 feet high). However, there are many other technologies available for diverting water, depending on the circumstances. For example, low dams (called barrages) stretched across a river serve the purpose well in many places. In others, pipes placed directly in the river with some minor headworks are common (for example, in the Mississippi for the New Orleans water supply). In other cases the water is extracted from the river by digging a very large-diameter well into the river bed and pumping the water out. This has the added benefit of serving as a sand filter for the water and is common in Europe, where the rivers were more polluted than in the United States at the end of last century when many of the European cities were building large municipal supplies (hence the term "French well").

The quality aspects of surface water sources have long been a major concern, hence in many cities the water sources chosen were located far away from sources of accidental pollution. This seems to have worked well in the case of New York City, with its protected catchments many miles away from major population centers and its carefully controlled land uses in the catchments. This is not the case in other cities, however, such as Boston, where suburban development now seriously compromises the quality of some of the nearby sources. For many industrial and agricultural uses this is not significant but it certainly is for most municipal water users. The potential (in some cases actual) contamination of the sources has led, in the most recent Safe Drinking Water Act, to the demand that all surface water sources be chlorinated before being delivered to the consumers.

For groundwater sources, the most familiar method of diversion is the modern pumped tubewell, so called because a tube is drilled into the ground and a pump is then lowered to pump the groundwater out. The pumps can be powered by electricity, diesel, natural gas, or even by humans and animals. There are at least 100,000 municipal water wells of this kind in operation daily in the United States, accounting for about half of the water used for irrigation. Wells up to 1,000 feet deep are in operation. The technology for the old open dug well (or "wishing well"), activated by a rope and bucket, was developed in antiquity. Jacob's well in Cairo, Egypt, was said to be 300 feet deep dug through solid rock. By the Middle Ages very elaborate human- and animal-powered pumps and other water-lifting devices were in everyday use. A glance at Agricola's *De re metallica,* originally published in Latin in 1556 and translated into English in 1912 by a leading mining engineer, Herbert Hoover, who later became president of the United States, will show the exquisite aesthetic of the old technologists.

Groundwater, which now provides one-half of all the water supplied in the United States, has always been popular for domestic uses because of the natural filtering of the water as it is pulled through the sandy layers of the soil. This natural sand filter had been spotted early by the Europeans in developing the French well mentioned above. As a source, however, groundwater also has some serious drawbacks. First, it tends to be high in the minerals dissolved in it on its journey through the ground; iron, calcium, magnesium, and fluorides are among the naturally occurring minerals that cause taste and odor problems. More recently, leachate from landfills, chemical storage and treatment lagoons, and other land uses such as septic tanks and agricultural chemicals and fertilizers have been showing up in the groundwater at noticeable levels.

Transport from source.   Once the source and the technology have been identified for procuring the water, it must be transported to the user. The technological solutions to this problem were highly developed millennia ago. For example, the Romans built an elaborate system of 11 aqueducts to supply the city of Rome with close to a million gallons of water per day from protected springs 30 to 50 miles away. In the ancient civilizations of the Middle East large unlined, and sometimes lined, canals were built to transfer municipal supply and irrigation waters. The Roman aqueducts were either open or closed depending upon the local conditions. The only thing that distinguishes these early works from present-day projects is the use of modern construction equipment and the replacement of some of the construction materials by plastic and steel. Nowadays some of the construction proceeds at a faster pace, but not necessarily so; it would probably take us longer to build the Roman aqueducts than it took them (an estimated 15 years) because of bureaucratic delays occasioned by environmental and other regulations.

The old systems, of course, relied on gravity to move the water; now we have lots of mechanically driven pumps to enable water to "flow uphill." Tunneling and large-diameter pipes along with pumps also allow the radical shortening of the transport distances.

Treatment before use. Most uses of water require some form of treatment before use. Even irrigation water needs to be desilted before being applied to the fields, and cooling water for industry and electric power plants may need to be stabilized bacteriologically to avoid biofouling of heat exchangers. Industrial process water often has to be of at least as high a quality as drinking water. Different sources of water contain different dissolved minerals, carry different levels of suspended material, and are of radically different bacteriological quality. Essentially there are only three basic types of treatment processes used in all of these cases: sedimentation, chemical precipitation, and filtration.

The sedimentation process is largely a physical process in which the water is held, or its rate of flow is slowed, so that suspended material will settle out due to gravitational forces. A huge array of technologies are available to achieve this, depending on the specific use and the required quality level. Silt ejectors in irrigation canals and sedimentation basins in municipal water treatment plants are examples of these technologies.

Two major processes are at work in chemical precipitation. The first is the addition of chemicals to transform dissolved materials into insoluble substances that will precipitate out of solution. The second is the addition of chemicals (e.g., alum) that form flocs; those suspended solids that are too small to settle out under gravity (in any reasonable length of time) are attracted to the flocs and settle out with them.

The filtration method was originally conceived as a purely physical process in which suspended material too small to settle out in the sedimentation process was removed by straining the water through some finely packed porous material, such as sand or charcoal. However, it was observed that biological and chemical reactions also took place on the filters. This led to the development of trickling filters for sewage treatment and activated carbon filters in water treatment for removal of dissolved chemicals. The first major breakthrough in the control of waterborne epidemics was the development and widespread adoption of sand filters for water treatment works at the end of the nineteenth century.

The final step in most treatment processes in the disinfection of the water to reduce the hazard of transmitting waterborne diseases. This step is often used even when the water is not for human consumption. It is possible to avoid problems of biofouling and corrosion by greatly reducing the levels of biological activity in the water while it is being used for industrial processes. For example, cooling water used in electric power plants or in air conditioner chilling units is treated this way.

Distribution systems. The distribution system is a major component of the capital and operating costs of a water supply system. As much as 30 percent of the total cost is embedded in the distribution system of pipes and balancing reservoirs.

The local transmission system is the source of the largest losses in the overall water supply system. Losses as high as 40 percent have often been reported, with 20 percent commonplace and even well-maintained systems losing about 7 to 10 percent. Distribution system losses are the first item to consider when faced with increased customer demands, since reduction of these losses can provide the cheapest source of "new" water.

Concerted efforts at leak reduction can also have tremendous impacts upon the viability of the current supply. For example, the City of Boston reduced the losses in its distribution system from more than 30 percent to less than 12 percent between 1978 and 1983 by an aggressive program of leak detection. Within a city, the capacity of the distribution system is typically set not by the size of the domestic or commercial demands but by the size of the maximum expected fire demand. This raises a series of difficult economic issues with regard to pricing.

Within the distribution system a series of local stores has to be constructed to balance the flow and pressure in the network. These range from the classic mushroom-shaped water towers to large underground storage ponds. The detention period of water in these tanks is now becoming a critical parameter in the disinfection strategies of the water suppliers. Long detention times (on the order of a few days) imply the need for a higher chlorine residual at the treatment plant to ensure the bacteriological quality of the water to the consumer, but at the same time this increases the levels of trihalomethanes (a group of chemicals, including chloroform, that are formed by the unintended action of the chlorine in oxidizing any organic material present in the water when sterilized using chlorine).

Treatment after use. In most industrialized nations waterborne waste collection is the norm. In many of the older cities the waste collection system was incorporated into the storm water drainage system. These "combined" sewers predominate in the United States and are the source of major spills of pollution to urban water bodies because of the instantaneous overloading of the sewer pipes during heavy rainstorms. The systems are designed to deal with these flow surges by use of appropriately located overflows: typically into the nearest water body, but also often into bypasses at the treatment plant. Separating sewers so that the storm drainage and the sanitary sewage can be treated separately is one of the contentious recommendations for meeting water quality standards in many older cities. Nationwide, separation of the sewers could cost as much as $100 billion.

As in the raw water treatment systems, there are very few different processes involved in wastewater treatment. A conventional partition is usually made at three stages. These are categorized as primary (where there is about 30 percent removal of the organic load and 70 percent of the suspended solids), secondary (about 80 percent removal of organic load and 95 percent of suspended solids), and tertiary (up to 100 percent removal).

In the primary stage, the raw sewage is usually passed through racks or screens to remove large objects (greater than 1 inch), and then passed through a comminutor to grind any remaining particles to 1/8 inch or smaller. Sewage passes through a grit chamber to remove the heavy sand and grit without removing the organic material, then goes into a large sedimentation tank where it is detained long enough to remove as much organic material (by settling) as possible without the whole process becoming anaerobic (which would cause nuisance odors and slow down the treatment process). After this the effluent, still containing maybe 65–70 percent of its original organic pollutant load, passes into a receiving body of water. The settled materials from the earlier steps, known as raw sludge (a highly odoriferous, unpleasant, and dangerous material which is about 96 percent water) must be further treated to stabilize it before ultimate disposal. This is usually accomplished using sludge drying beds, where thin layers of sludge are allowed to dry under atmospheric conditions until it can be carried away and landfilled.

The secondary process is usually an add-on to the primary. The effluent is passed through a treatment process known as biological activation to remove most of the remaining suspended solids and much of the dissolved organic material. In this process, the biological decay of the wastes is enhanced by adding more bacteria and stimulating their growth, using the wastes in the water as the food for bacterial growth. The goal is to encourage the right kind of aerobic bacteria by providing them with ideal temperature and oxygen concentrations to maximize their reproduction, and hence the oxidation of the wastes. Trickling filters and activated sludge units used in many sewage treatment plants are based on these principles.

Secondary treatment is usually accompanied by some form of advanced sludge handling techniques. These typically could be anaerobic digestion followed by mechanical sludge drying. The digested sludge is a safe and stable material that has been used in many parts of the world as fertilizer or mulch. The undigested sludge has been used as fertilizer for trees or other field crops not destined for direct consumption by humans. Sludge produced by a secondary treatment plant adds as much as 40 percent to the quantity produced by the primary process. Sludge disposal often only becomes a major problem when a municipality upgrades its facilities to secondary treatment. The current federal water quality laws require at least secondary treatment for all but a handful of coastal cities (Los Angeles is the largest city with such a waiver).

Occasionally a third stage (tertiary treatment) is added to achieve additional purification by the removal of phosphorus, nitrates, and other contaminants that are not removed by primary and secondary treatments. Tertiary treatment is typically much more expensive than both primary and secondary.

In the water policy area, the improvement of water quality is usually premised on technical fixes. For example, in the area of groundwater contamination most engineers believe that it should be possible to create leakproof landfills despite decades of evidence to the contrary. Moreover, if the landfill should indeed fail they tend to believe that the problem can be solved by yet another engineering trick. At some basic level of analysis all of the support systems used by society can be considered technical fixes. The wish to continue adding technical fixes is an act of faith by most engineers.

Technological Imperatives and Political and Social Biases

From the point of view of water policy it is important to establish which factors are inherent in the technology and cannot be easily changed or adapted to changing social needs. The most important imperative from this point of view is that almost all the technical options for water control exhibit large economies of scale, i.e., the unit cost of supply or treatment becomes less as the size of the unit is increased. This leads to building infrastructure as large as possible, which in turn leads to large amounts of overcapacity, which then encourages excessive use to take advantage of this excess capacity.

The second imperative that influences policy is that of the difficulty of measuring and monitoring ambient water quantity, quality, and environmental impacts. This leads to preferences for end-of-the-pipe regulation rather than broader regulation of process change, renovation, and recycling, and leads to emphasis on end-of-the-pipe effluent standards rather than ambient environment standards for quality.

The technology also has a bias toward capital expenditures rather than operating and maintenance expenditures: toward building large capital-intensive projects that require little further maintenance once constructed. These types of investments are easier to control and operate and are less liable to operator failure than small labor- and management-intensive schemes. This bias has, however, inhibited innovative developments in many areas of water control, supply, and treatment; renovation and recycling of wastewater is one area where its effect has been felt most strongly. The bias toward capital-intensive development was greatly encouraged by the EPA's construction grant program, which provided as much as a 75 percent federal subsidy on capital expenditures and no subsidy for operation and maintenance expenditures. In this particular case the level of the subsidy was decisive in choosing the technology, not economic and engineering considerations.

Finally, the choice of technology can have major social implications. This is most clearly observed in the construction of irrigation projects. If the choice is to build dams and canals, certain riparian social groups are greatly benefited at the expense of everyone else. This is acceptable if it is agreed that these groups should benefit, but often the irrigation canals flow through the lands of the big rather than the small farmers. If, on the other hand, the choice is groundwater development, this can be made available (more or less exactly) to the targeted groups. Another example of technology's social consequences is the provision of water mains and sewers to previously unserved areas. Areas that previously could only support low-density development based on local wells and septic tanks are now available for intensive land speculation and development. Again, the EPA's interceptor sewer grant program encouraged these patterns of growth.

What Parts of Technology Choice Can Be Safely Left to Engineers?

Because of the technical nature of water control many of the decisions regarding the design and implementation of water infrastructure are left solely to engineers. Unfortunately, for many aspects of the decisions involved engineers are no better qualified than others to make the choice. Analysts who examined the impact of sewer planning on land use (Tabors, Shapiro, and Rogers, 1976) concluded that there were sets of design parameters that one could confidently leave to the

Table 5.1
Parameters of Collection System Design That Can Be Left to Engineers

Parameter	Physical Design Considerations	Potential Input from Planners/ Informed Citizens
Basic land use/demographic parameters	Service area Land use, densities Population growth	Major
Natural features	Topography (slopes) Soil types, water table	Some, as source of data
Capacity requirements	Per capita flow Peak-to-average flow	Some, in developing water use projections
Engineering design criteria	1. Pipe characteristics: Manning's n 2. Joint materials 3. Minimum pipe diameter 4. Minimum velocity 5. Maximum velocity 6. Minimum and maximum slopes	Little

Source: Tabors, Shapiro, and Rogers (1976), p. 41.

professional judgment of the engineers and others that could not (see table 5.1). In the case of sewer design, two overwhelmingly important parameters that are routinely left to engineering designers are the size of the population to be served by the sewer and the service area. Indeed, the design capacity is very sensitive to these parameters. Moreover, the provision of excess capacity in sewer lines has a marked tendency to induce population movement to the newly sewered area and, hence, to make the population forecast self-fulfilling.

In other cases we see engineers choosing interest rates to discount future benefits and costs that are so unrealistic that huge projects are justified. As part of the same problem we see excessively long design periods that determine the technology to be used into the distant future, hindering society from taking advantage of modern inventions and developments. But for federal water projects the discount rate is set by law and may be set too high and give the opposite effects: projects too small or not justified at all.

Economic, Financial, and Public Expenditure Imperatives

Economic Concepts Underlying Federal Water Policy

In water policy two fundamental economic concepts of efficiency and equity play a major role. For the first time in any federal government activity, the Flood Control Act of 1936 introduced the requirements of a formal benefit-cost analysis in justifying water projects. This was enacted even before economists developed a detailed methodology for measuring the benefits accruing to public sector investments.

The economic imperatives of water are inherent in its nature as a fugitive resource that is often treated as a public good. External effects occasioned by water use also invoke important economic imperatives. For example, property rights influence the choice of ways to price water to pay for the public supply of it. To appreciate water policy options and how they are evaluated, it is necessary to understand how economics is used, and misused, in the water area. Economics is currently the most important dimension of water policy, but politics ultimately controls the outcome, even when the discussion is framed in economic terms.

Everybody seems to have some instruction in economic theories and practice. For example, most people would agree that if the price of water increases, consumers will use less of it. There is a *willingness to pay* that is not merely an abstract economic concept based on theories of private property but is a reliable behavioral trait of consumers everywhere. It is, therefore, a good basis for assessing the economic demand for water. If human nature acts the same in widely differing locations, the natural environment can also be expected to respond to human modification in similar ways in different locations. Thus, the most accessible water for human use is the cheapest to develop, the next most accessible costs a little more, the next a little bit more, and so on. Very soon one can observe an increasing *marginal cost* curve that depends upon nature and not upon theory. Between these two observed phenomena—of human nature and the natural environment—a good set of practical rules can be derived for helping decide how much to invest in developing a water supply.

To decide whether to spend money on irrigation or flood control systems, or on investments outside the water sector, these simple behavioral rules need to be bolstered with additional assumptions and theory. Much of the economic literature is devoted to explaining how to decide which are the most plausible and least constraining assumptions and theoretical constructs. I believe that it is useful at this stage to look at the fundamental economic theories of resource management and see how they apply specifically to water.

Valuation of Water

Allocation of water among the myriad conflicting potential uses presents a major task to government, which must take responsibility in some degree for regulating access to water. It is difficult to assign unambiguous economic values to many uses, and hence these may be implicitly overvalued, undervalued, or completely ignored in the decision-making process. Many of the problems of valuing water stem from market failures; in particular, the existence of externalities and the lack of mobility of resources make finding the correct price by which to value the resource quite difficult.

In a perfectly functioning economy envisaged by the classical economic model, price equals value, and the cost of providing a good, after allowing for payments to all of its factors of production, will precisely equal its market price. As a result of this elegant solution one only has to establish cost to establish value. Unfortunately, many water resources planners forget that equating cost with value only holds true in a perfectly functioning market economy. In all other cases (that is, almost all cases) care must be taken not to confuse cost with value.

What then is the value of water? The answer appears to depend upon to whom and for which use. Drinking water is obviously valuable and becomes increasingly so as the amount available decreases. A glass of water could be infinitely valuable to a person dying of thirst in the desert but not very valuable to a person living alone on the banks of a pristine river. In the second case the person would only be willing to pay the cost of somebody going to the river and fetching the water for her. Alternatively, she could go and take it herself. So the value in this case is the cost of saving herself the trip. If there were also a farmer irrigating land alongside the river, how much would the water be worth to him? If there is enough water in the river so that the woman can have as much as she can drink, just at the cost of obtaining it, then obviously the farmer can also take as much as he wants at the cost of obtaining it, and would also value the water at that cost. So far, so good; cost equals value.

Unfortunately, such bucolic settings no longer exist in the modern world. Typically there are many users of the resource, and at some point the use by one user will start to interfere with the use by another. At that point the water is said to have an *opportunity cost* since there is a loss of the opportunity to use the water by one user. The cost to the affected user is the amount she values the water. At this point the value of the water should reflect the willingness to pay of the user who is losing water. If for some reason the consumer has to cut her consumption of drinking water, then the opportunity cost to society is her willingness to pay for it. If the allocation of the water shortage affected irrigation, then the relevant opportunity cost would be the farmer's willingness to pay for it.

Now, if the question of how to allocate the water were left to an outside party, for instance the U.S. Army Corps of Engineers, then that party might ask how society would best benefit from the allocation. One way of answering this question is to apply the logic of social choice theory embodied in modern economics, which implies (all other things being equal) allocating the water to the use with the highest value. Another way is to apply the Principles and Guidelines developed by the U.S. Water Resources Council (1983). These guidelines were specifically developed to take into account social preferences concerning water and land developments.

Establishing the willingness to pay for commodities by consumers is a fairly well developed field in economics and can be easily adapted in many water conflict situations to establish estimates of the opportunity cost of water. Unfortunately, many economic studies only reflect the actual costs of obtaining the water, not the opportunity costs. If there were well-established markets for water then the market price would itself reflect the opportunity costs. However, for many uses of water free markets do not exist and one is left having to estimate the opportunity cost in indirect ways.

The opportunity cost of water is zero only when there is no shortage. In evaluating water investments it is therefore important to remember that *the value of water to a user is the cost of obtaining the water plus the opportunity cost.* Ignoring this will lead to undervaluation, to failures to invest, and to serious misallocations among users. The opportunity cost concept also applies to issues of water and environmental quality.

Public Goods and Externalities

Some analysts claim that the resources that make up the environment are unsuitable for private ownership because they lack the "excludability property." In other words, it is not practical to exclude people from using the resource because it is either physically impossible to limit access (breathing the air) or very expensive or cumbersome. Improved water quality in lakes and rivers, the guidance of navigation lights, public beaches, security from flood damage, and many other investments that improve the quality of the environment have this property to a greater or lesser extent, although at one time or another in the history of the United States they have indeed been considered private property. There is little incentive to provide services or own property from which other people cannot be excluded. Everyone's property is no one's property.

Nonexcludability is not the only thing, however, that makes environmental resources different from the classic privately owned resources. In addition, they have the property of mutually interfering usage. Individuals take the valuable commodities of clean air and water from the same environment into which they then dump wastes that interfere with the use of the no-longer-clean air and water by themselves and others. In economic parlance these conditions are referred to as externalities. It is both nonexcludability and externality that make water an inherently difficult resource to manage.

Economists have arrived at four criteria for judging such policy decisions. The first two relate to "welfare" or "satisfaction" in utility terms; the second two relate to the productivity of the economy in monetary terms. The presentation here uses the framework suggested by Dorfman and Dorfman (1972).

Economic Concepts of Efficiency

The institution of private property and the other economic institutions in the United States have evolved together in ways that tend to promote the efficient use of things that are privately owned. A corollary to this would be that if a resource is not privately owned, then the developed institutions do not work well in promoting its efficient use.

The economic concept of efficiency differs radically from the engineering and scientific concept, yet engineers as well as economists are required to analyze, promote, and implement water resources policies efficiently. It is therefore important to clarify the use of this word. Most irrigation engineers attempt to find the "most efficient" use of the water—meaning a system in which there is a minimum of waste and where the highest percentage of water reaches the roots of the crops. An economist looks at the same set of conditions but considers water as just one of many resource inputs that should be used to achieve the best economic (or "social") outcome. For the economist, it is not at all obvious that the most important goal is to avoid wasting a drop of water, because water may not be the limiting resource in achieving the greater objective of using all resources wisely. Economic efficiency, then, carries with it some notion of an objective that is broader than the use of one input. But many of the water policies promoted by the U.S. government (at home and abroad) are guided by the engineering concept of efficiency, saving some water but misallocating very large amounts of public monies.

Broad utility criterion: Pareto optimality. Pareto optimality is usually described as an economic equilibrium where a reallocation of resources among members of society is not possible without making at least one member of that society worse off (i.e., less satisfied than he was with his previous allocation). Pareto optimality can only measure which outcome an individual prefers in any given choice between a pair. The tradeoff between dollars and utility is left to each individual to decide; hence, monetary transactions are perfectly appropriate as measures of utility and satisfaction. This type of analysis could also be applied to groups of individuals if it were possible to assess their group preferences, but in any realistic setting the assessment of Pareto optimality is practically impossible.

Sharp utility criterion: social welfare.   If it were possible to devise a community "welfare function" as some function of the welfare of its individual members, then it would be possible to improve upon the broad utility criterion. It would have to take into account the relative merits of each member in defining the welfare of the whole, since the equitable distribution of the utility among the members is a prerequisite of a welfare function. The welfare function will help discriminate between the infinite number of Pareto efficient combinations by choosing the Pareto optimal solution that maximizes the social welfare function.

The elegance of a community welfare solution is diminished by the fact that no entirely satisfactory social welfare function dependent on individual utilities has ever been constructed. Nonetheless, the economic literature spends a great deal of time discussing the properties of this hypothetical construct—tending mainly to erode the credibility of economics among policymakers, even though economists have something profound to offer environmental policymakers.

Broad productivity criterion.   In the practical world of goods and services the concept of Pareto optimality has an analogue, the concept of productivity efficiency. An economy is said to be productively efficient if it produces as much of every good and service as is possible, given the level of the outputs of all goods and services and the level of resources used as inputs. For example, consider an aquatic economy that produces just two commodities: rice grown under irrigation and water-based recreation. For a given amount of resources, one could have either more recreation or more rice production, and be efficient in either case. An economy can be productively efficient, however, without being Pareto optimal. This means that at several points along a spectrum—the "production-possibility frontier"—the economy is productively efficient, but it is the wrong combination of commodities that are being efficiently produced. This is not an unusual phenomenon in many centrally planned economies. Despite this drawback, the broad productivity criterion is attractive because it is computable in many circumstances where it is not possible to compute or even conceive of Pareto optimality. The criterion does not, however, help choose between efficient points; it is as far as we can go without making major new assumptions about how the economy "ought" to work.

Sharp productivity criterion. Neoclassical economic theory states that the relative desirability of private goods is reflected in their prices. Accordingly, the best point to choose on a production-possibility frontier is the point where the value of goods and services produced is greatest. This is referred to as the GNP criterion—the point where gross national product is at a maximum. In terms of the example given above, assuming that irrigated rice is worth $1 per pound and recreation is worth $10 per day, then the equation for GNP for this two-commodity economy would be:

GNP = $1 × number of pounds of rice produced
      + $10 × number of days of recreation provided.

The obvious choice to maximize the value of the GNP is that combination of goods produced at the point where the value of the GNP is as large as possible, given the requirement of staying within the production-possibility frontier.

From a practical point of view this criterion is the most attractive. It is straightforward to compute and captures the ideas of its predecessors without the drawbacks of most hypotheticals. However, despite its appeal and widespread use in benefit-cost analysis, it has some serious flaws in application.

Summary of economic efficiency criteria. Even though the economic literature emphasizes efficiency and equity as the two goals of economic endeavor, operational definitions of equity are much harder to pinpoint. Of the four criteria discussed above (broad utility, social welfare, productivity efficiency, the GNP criterion) only the social welfare model deals directly with the issue of equity. Unfortunately, that is also the one that is most hypothetical and least operationally useful, leaving the policymaker with some good tools for social analysis that are unfortunately not completely helpful for public decision making in democratic societies.

A resource reallocation decision is said to be Pareto efficient if the decision can improve one individual's well-being without decreasing the well-being of anyone else. Most allocation decisions in the public domain involving water resources appear to make some people better off and some worse off and, therefore, cannot be evaluated using the simple Pareto concept. Economists have therefore developed the compensation principle to extend the relevance of Pareto optimality. The Kaldor-Hicks compensation criterion is widely used; under it, a reallocation is a "potential" Pareto optimal if the gainers would be able to fully compensate the losers and still be better off themselves. For strict Pareto optimality the compensation would actually have to be paid.

In real cases, however, there is typically no easy way to make the compensatory payments. For example, the beneficiaries of a large federally financed water project could be charged for the water supplied, and the revenues generated could be used to compensate those who lost access to land and water as a result of the project. However, the beneficiaries are usually a small, clearly defined group receiving substantial benefits, whereas the losers tend to be a large number of widely scattered people each suffering small damages. How does one get them a priori to agree on equitable compensation?

The compensation test can be used as an analytical tool even though compensation will not actually be paid. Taking the value of economic goods and services gained or lost, one compares the gains of the gainers to the losses of the losers. If gains outweigh losses, the conclusion is that the overall welfare is advanced. This is the conceptual basis of benefit–cost analysis included in the Flood Control Act of 1936: "The Federal Government should improve or participate in the improvement of navigable waters or their tributaries including watersheds thereof, for flood control purposes *if the benefits to whomsoever they may accrue are in excess of the estimated costs*" (emphasis added; Holmes, 1972, p. 19).

The usefulness of performance goals expressed in compensation terms, however, becomes clearer when contrasted with goals that are so broad as to be virtually meaningless; for example, those expressed by the Natural Resources Planning Board in 1939: "That is, in order of (1) the greatest good to the greatest number of people, (2) the emergency necessities of the nation and (3) the social, economic and cultural advancement of the people of the United States" (Ex. Order 8248).

Even though the Pareto discussion is carried out in terms of the preferences of groups, the actual measurements used in applying both the compensation principle and benefit–cost analysis are those based upon the GNP criterion. Under this, everyone's welfare is counted equally, in direct proportion to the goods and services gained or lost, measured in market prices. Since the demand for goods and services is known to depend upon income, the assumption is that the current distribution of income is acceptable and can be used as the basis for further allocation of goods and services. The compensation principle has been widely criticized as a basis for public policy decisions on this score.

The Principles and Guidelines (P and G) mentioned above have attempted to codify the economic concepts discussed in this section to evaluate federally funded water projects. The P and G represent a careful attempt to reflect the best economic and environmental thinking in producing a methodology that can be used by the federal establishment. The reality, however, is far from the nobility of the conceptual apparatus, and the applications of the P and G still need further development. In particular, the P and G need to be made applicable to regulatory action as well as project development. This would then require the EPA to use the P and G to evaluate its regulatory actions. Ideally this could lead to a major shift in the philosophy of regulations currently employed.

Demand for Water and Water-Based Amenities

The "demand" for water is a factor of economics, specifically the quantities of a commodity that consumers are willing to purchase at various prices. In market economies resource allocation and distribution problems are solved simultaneously by the price mechanism. The demand for a certain resource is brought into balance with available supply at some price level: as demand increases the price rises and some consumers use less than they would have at the lower price; if demand declines, or if new cheaper sources of the resource become available, the price drops.

In the strictest sense, the concept of a demand schedule applies only to consumer purchases. Resources and material used as inputs in the production of some other commodity that is purchased directly by consumers (for example, irrigation water used for growing food) reflect "derived demand," since call for them derives from the final demand of the consumer product that they are used to create.

Unfortunately, in most public discussions of water the word "need" is often mistakenly used for demand. But it makes no sense to project quantities of water needed without specifying how much people would be willing to pay for these quantities. Demand, therefore, must be specified by two numbers: quantity and price. In planning future water development, current water use rates are usually put on a per capita basis and projected according to estimated population increases. This leads to projections of excessive water "needs." Figure 6.1 shows some of the projections for future water demand made for various planning agencies over the past 17 years. Note the tremendously large range of predictions, and note that the actual water consumption is below all of the forecasts. The decade from 1980 to 1990 saw increasing prices for water in most parts of the country with a concomitant decline in demand, exactly as economic theory predicts. This did not account for all of the decline, however, since stringent water quality standards led to substantial industrial recycling, and reduction in demand for agricultural products led to a decline in irrigation use.

Demand curves. The concept of demand curves was first put forward by the French engineer Jules Dupuit (1804–1866), who among other things invented the concept of consumer surplus and did fundamental research in hydraulics, while he was the Inspector General of Roads and Bridges in France. The concept of the demand curve is very simple; the amount of a commodity that would be consumed is estimated over a range of prices based on observations of the same consumer over time as the price for the commodity has changed. However, the concept immediately becomes complex when applied to a commodity like water. First, there are many uses for water; it behaves like several different commodities at the same time. Second, depending on how water is used, it may also belong to one of several markets.

While demand curves originally applied to only one market for one good and for one consumer, demand curves for groups of consumers with similar characteristics were established early, as were demand curves for one commodity used in several markets. In the southwestern part of the United States (figure 6.2), for example, consumers are willing to pay very high prices for residential and commercial water ($300 to $600 per acre-foot); industrial users are willing to pay considerably less ($120 to $300 per acre-foot); and agricultural users pay the least (well below $100 per acre-foot for crops other than flowers, fruit, and vegetables). Only small quantities are demanded at high prices and large quantities are demanded at low prices.

Supply curves. Another important concept in understanding the economics of water resources is that of a supply curve. The supply curve is created by arranging, in ascending order of cost, the supplies

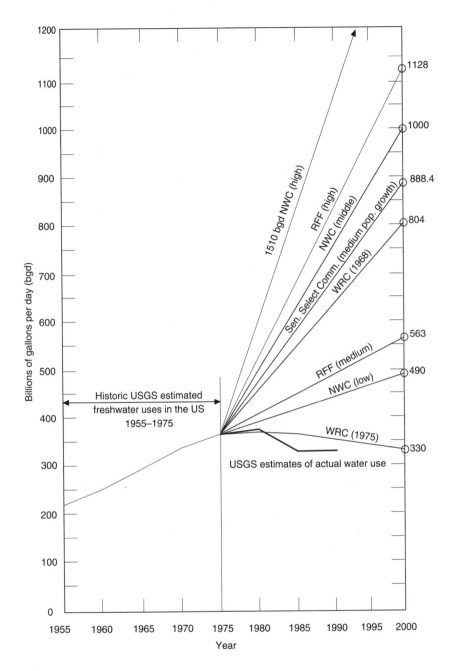

Figure 6.1
Historic and projected freshwater withdrawals, 1955–2000. The estimates are by the National Water Commission (1973); Resources for the Future (published in Wollman and Bonem, 1971); the Senate Select Committee on National Water Resources (published in U.S. Congress, 1961); and the U.S. Water Resources Council (1968, 1977). (Source: Congressional Research Service, 1980.)

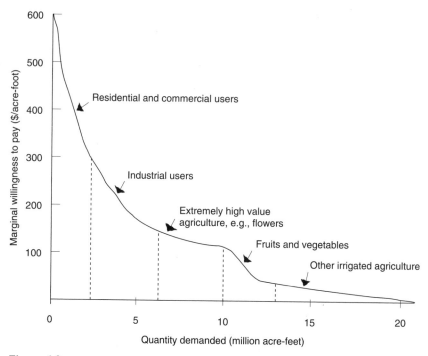

Figure 6.2
A typical water demand curve.

of various magnitudes available to water consumers in a given market area (figure 6.3). In the southwestern United States, the cheapest water, but in limited quantities, is available from local surface water diversion. The next least expensive water is groundwater, which is available in much larger quantities. The next least expensive water is from interbasin transfers, followed by wastewater reclamation (also limited in quantity). Finally, desalination is available in almost infinite quantities at a very high price.

Market clearing.   One of the great conceptual breakthroughs in economics is the idea that the most efficient economic solutions occur when the amount a consumer is willing to pay matches the supply price; at that point the market is said to "clear." This concept holds as long as a number of other conditions also exist, but even in its simplest form it does provide planners and policymakers with a set of simple tools drawn from engineering and physics (supply curves) and observed behavior (demand curves) that give good indications of

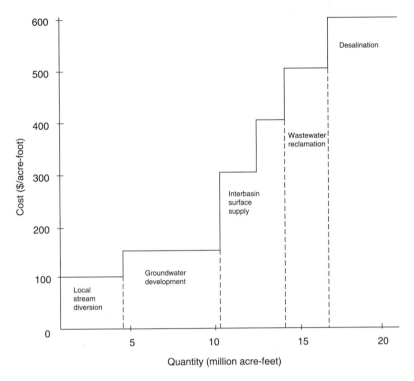

Figure 6.3
A typical aggregated water supply curve.

efficient economic solutions. By superimposing the supply curve on the demand curve one can identify the price and the quantities at which the market would clear.

This discussion highlights the role of prices in encouraging or discouraging excessive use of water. If, as is often the case, the price of water is held unrealistically low, the demand for water grows very large. As consumption grows the costs of supply increase, which often results in an increase in the public subsidy that frequently exists, rather than in an increase in the cost to consumers.

Due to market restrictions on the transfer of water rights it is possible to have multiple markets for the same commodity in the same region. In such cases, prices cease to give the correct rationing signals. The water market in California is a good example of this. Moreover, in markets that are heavily regulated, such as municipal utilities, the market will not clear where the supply curve intersects the demand curve because the price is set either too high or too low by the regulators. If the price is set below the marginal cost, then consumers must be subsidized. These are called "deadweight" losses. If the price is set above the marginal cost, then the consumers will actually use less than they would have were it set at the economically efficient level. This less than optimal consumption means a "welfare loss" for society; marginal cost pricing is the economic prescription for tariff setting.

Price and income elasticity.   Ernst Engel (1821–1896) was the first person to formulate the concept of elasticity as it is now used. He noticed that persons in higher income classes in Germany spent a smaller proportion of their income on food than people in the lower income groups. He defined the income elasticity of food as the ratio of the percentage change in quantity consumed to the percentage change in income. Engel found that this ratio was less than 1; in other words, the relative amount of expenditures for food declines as income increases.

If the absolute value of the elasticity is less than 1, the consumption is said to be "inelastic," and if above 1 it is said to be "elastic." Elasticities can be measured for almost any two sets of quantities that vary simultaneously, but the important ones from the point of view of water are those of price and income. A price elasticity for municipal water supply of −0.7 implies that for a 10 percent price increase the demand for water would decrease 7 percent. Even though this demand is said to be "price inelastic," it does not mean that pricing cannot be used as a rationing tool for municipal water. (If water were price elastic, the impact of pricing would of course be even greater.)

Water is also typically income inelastic (the demand increases with increases of income but less rapidly). The joint effect of price and income effects needs to be taken into account in making forecasts of future demand. For example, the U.S. Department of Commerce (1989) forecast that the population by the year 2000 will be 267 million (that is, growing at about 1 percent per annum from 1980). How large will the demand for water be? Simply assuming water use proportional to the population growth rate, it would increase to 22 percent above the 1980 level. If the forecaster wanted to include the effects of increasing income levels (estimated to be growing at 2 percent per annum over that period) he would need to know the income elasticity of water. This is typically about 0.5. Factoring this into the forecast would lead to a projected 49 percent increase in water demands by 2000. So far this assumes that price remains constant. What if the forecaster allowed for a price increase (at about 3.5 percent per annum) and assumed a −0.5 price elasticity? Then the demand drops to only a 5 percent increase over this 20-year period.

According to this simple example the United States could have a water "crisis"—the 49 percent increase—or just a modest increase in demand—the 5 percent increase. Which forecast should be used? The answer depends upon whether a prescriptive or a descriptive stance is taken. The choice is up to governmental policymakers. The "needs" forecasts do not reflect the restraining effects of price. If the regulators leave water sellers free to make water prices that more nearly represent the marginal cost of supply, and if realistic pricing policies are pursued in cases where the supply has to be controlled by government, then the forecast crisis will never take place.

Measurement of Benefits

Conceptual issues. It is easy to say that the benefits must exceed the costs. But how can we measure benefits? In a perfectly functioning market, price is a measure of value. Therefore, the total benefit of consumption can be simply measured by the sum of the price times the quantity of the commodity consumed, and in fact this is a perfectly good way of estimating benefits for water projects. But what of outputs for which there is no traditional market (e.g., water quality and aesthetics), or for which there is a market but entry is constrained, or for which there are large economies of scale (e.g., irrigation water, navigation, and hydroelectric power), or cases where the project creates a unique market (e.g., water-based recreation in an arid area)? These are hard cases that have nonetheless been satisfactorily resolved conceptually by the economics profession. In many cases, robust methods have been developed for operational measurement of the benefits.

A comprehensive review of all the different types of benefits of water projects is given by Desvouges and Smith (1983) and shown in figure 6.4. They split the benefits into two major categories of "current user values" and "intrinsic or option values." The current user values themselves are split into two major categories of "direct" use and "indirect" use. Direct uses represent the obvious benefits of municipal, agricultural, industrial, commercial, recreational, and hydro-

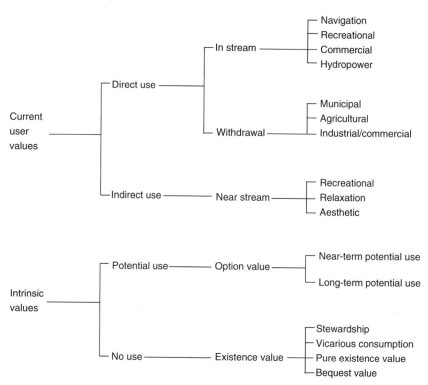

Figure 6.4

Types of values associated with water resources projects. (Source: Desvouges and Smith, 1983.)

power use that feature in the conventional analyses of water projects. The indirect uses include recreational, relaxation, and aesthetic benefits, not usually included in conventional studies.

Desvouges and Smith's other major category, intrinsic values, have until now almost never been considered in project analysis. They include benefits that would accrue to potential consumers sometime in the future. Finally, there are the "no use" benefits based upon the existence value of the project itself, associated simply with the knowledge that the project exists (or does not exist), and with good stewardship of the earth's resources. The intrinsic existence benefits are the least well defined, but individuals are willing to pay for these benefits (or to avoid disbenefits in these areas) and hence they are just as real and important to evaluating a water investment or management decision as the more conventional benefits.

Since most of these "intrinsic," "option," or "existence" values tend to be preservationist or conservationist, they show up as negative benefits for most water decisions that lead to significant changes in the environment. This means that they are not popular with proponents of water development. However, if people are willing to pay for them, strict economic logic dictates that they must be considered even when they are difficult to measure. This is especially important because their neglect has generated much unhappiness about how current water policy is implemented.

Practical approaches.   In estimating benefits for water development the easiest uses to deal with are those that have well-established markets and firm monetary valuations for the outputs. Benefits from municipal, commercial, industrial, hydropower, and irrigation uses are relatively straightforward in comparison with benefits from navigation, recreation, and aesthetics. At the basic theoretical level, however, the approach to estimating each of the benefits is more or less the same. Essentially, it involves looking at the demand curves for municipal water supply, then estimating the change in the total willingness to pay for a change in supply. In the case where the demand curve is not known, the alternative cost of consuming the commodity can be used as the basis for assessing the benefits. For example, road or rail can be used as the alternative cost of transporting goods by inland navigation. The total willingness to pay is measured by the area under the demand curve; the benefit-cost criterion shows that the efficiency goal is the same as "maximizing the net willingness to pay."

The details on estimating the intrinsic values can be found in Fisher and Raucher (1984), Stavins and Willey (1983), and Meta Systems, Inc. (1985). Intrinsic benefits (such as preservation and conservation), in the cases where they have been estimated, appear to be of about the same order of magnitude as current user benefits. Hence they should no longer be dismissed out of hand by agencies promoting projects, even though they are likely to end up as negative benefits in the actual benefit-cost analysis.

The improvement of the methodology of intrinsic benefit measurement is of great importance for further development of the nation's water resources infrastructure. From a purely philosophical point of view, the incorporation of these benefits makes the rational economic model of resources management much more acceptable in the policy arena.

Application to Water Resources Management

Among all resources, water has probably received the most attention from the economics profession. Starting with the Flood Control Act of 1936, the "Green Book" in 1950, and Circular A-47 of the Bureau of the Budget issued in 1952, a series of studies led to the formulation of specific economic criteria to be applied to proposed federal water projects. They epitomized the federal concern with economic efficiency in government. This concern was stimulated by the general perception that water projects were pork barrels and otherwise unjustifiable. Yet economic analysis of the benefit-cost kind, in the hands of skillful practitioners in the Corps of Engineers or the Bureau of Reclamation, can also be used to justify many (some would say all) poor projects that are authorized for federal investigation.

Against this background two studies were carried out at Harvard University starting in the late 1950s. The first resulted in a book (Eckstein, 1958) that laid out the classical economic theory underlying the use of benefit-cost analysis for water resource planning and analysis. The book was notable for the fact that it was able to demonstrate that classical economic theory could handle all the complexities of practical problems of water. Unfortunately, the elegance of both the derivation of the first-order conditions and the mathematics hid a set of assumptions that some argued were fatal to the original premise. For example, it was necessary to assume that the marginal utility of income for all individuals is the same. For this to hold it would be necessary for the distribution of income in society to be optimal; in fact, the very goal of many projects is to redistribute income. Implicit in the model were also a set of assumptions about the consumers, the producers, and the structure of the market in which the water project functions. All of these assumptions are of varying levels of credibility.

Assumptions about consumers:

1. Rational consumers act consistently on preferences.
2. Successive units of a commodity add less and less to a consumer's satisfaction and hence add less and less to his utility (diminishing marginal utility of consumption).

3. Preferences must be independent of purchases by others (no keeping up with the Joneses).

Assumptions about producers:

1. Producers must pursue the principle of profit maximization rationally.
2. Production processes must not be such that successive units of production are cheaper (decreasing returns to scale).
3. There is no physical interdependence among production processes (no externalities).

Assumptions about the structure of the market:

1. Markets must be "perfect."
   i. Producers and consumers must have complete information.
   ii. Both producers and consumers must be small relative to the size of the market.
2. The resultant distribution of income is appropriate.
3. Labor, capital, and other resources are relatively free to move from location to location.
4. There is full employment of labor and resources.

Many of the above assumptions are reasonable "enough" (the assumptions about the consumers), or they are not reasonable but there is little that we can do about them (complete information about the market). There are a few assumptions about water policy, however, that particularly stand out as being too strong, or simply incorrect:

• *Increasing* returns to scale on the production side are prevalent in water projects. Inland waterways and municipal water and wastewater services, for example, are natural monopolies because of the large economies of scale in the provision of the infrastructure. Most water-related investments tend to be very large to take advantage of these economies of scale.
• Physical interdependence among production processes is inherent in many water activities. The externalities are experienced in the spatial sense of both water quantity and water quality between upstream and downstream users, in a temporal sense between different seasonal releases of stored water, common pool effects on groundwater, and the export of pollution.
• The income distribution assumption. Many water projects are specifically aimed at changing the distribution of income.
• Producers are not small relative to the market. When the federal government is involved it is often as the only producer in the market. Moreover, the water supplied will make large changes in the local price of water, undermining the assumption of marginality inherent in the benefit measurements.

• Resources are not necessarily mobile. In the United States, capital resources are relatively mobile but labor resources are not. Pockets of poverty and unemployment still exist and many water projects (like the TVA) were originally designed to address these problems. In addition restricted water rights often impede the ease of transfer of water from one use to another.

A second study at Harvard (Maass et al., 1962) attempted to remedy some of the obvious problems with the Eckstein study. Headed by a political scientist with the assistance of economists, engineers, hydrologists, and operations researchers, it took a broader approach because of its interdisciplinary organization. This study asserted that "the maximization of national welfare" was the goal of water development and encompassed two distinct objectives: economic efficiency, measured by the aggregate net willingness to pay, and income distribution to some target groups. Apart from the insistence upon equity being kept separate from economic efficiency and its highlighting of the importance of the means to be used to achieve this income distribution, the second study did little to remedy the market failure issues assumed away in the earlier study by Eckstein. It paid a great deal of attention, however, to measuring the benefits of water projects.

Contemporary with the Harvard studies there were several others, notably those by Hirshleifer, De Haven, and Milliman (1960), that presented a "Chicago School" riposte to the Harvard studies, and a series of publications by Kneese (1964) and his colleagues from Resources for the Future. While Hirshleifer et al. emphasized the use of market measures to assess benefits and the use of market rates of discount instead of the "social rate of discount" promoted by Harvard, Kneese and his colleagues researched pricing mechanisms for efficient allocation of water quality. But neither added new elements to the basic economic analysis, nor resolved its problems. The only major improvements in the economic approach have come from development economics, where explicit consideration of market failure has been successively incorporated into the economics of resource management. Little and Mirrlees (1974), Dasgupta, Marglin, and Sen (1972), Mishan (1976), and UNIDO (1978) showed practically how to assess social rates of discount and shadow prices on constrained and underutilized resources.

Benefit-cost analysis is now a widely accepted requirement in most federal water agencies when deciding whether a particular water project should be implemented, although most other federal agencies, accounting for 90 percent of discretionary federal spending, do not make use of benefit-cost analysis in their decision making. The EPA is foremost among the water regulatory agencies that do not employ such techniques. Agency personnel do not appear to be bothered by theoretical economic problems, and most are quite skillful at making the benefit-cost analysis work to the advantage of the agency's goals. For examples of such creative uses of benefit-cost analysis by the water agencies see the paper by Sander (1985), which shows large discrepancies between agency benefit-cost ratios and those of the Office of Management and Budget's analyses of the same projects. Some projects that showed a 2.5 to 1 benefit-cost ratio in the agency analysis only showed a 0.3 to 1 ratio in the OMB's analysis.

Cost sharing in federal programs.   Many observers hold that a key accomplishment of the Reagan administration was to insist on higher levels of cost sharing by the beneficiaries in federal water resources projects. Cost sharing has always been a feature of federal water projects but compliance has not been very good. In most cases there is a large discrepancy between the nominal cost share and the historical performance. Cost sharing has had a major impact on the number of studies and projects. The biggest constraint is *up-front financing* of the nonfederal cost share. In the past, nonfederal sponsors could pay off their share over the lifetime of the project. Not anymore; it is a serious constraint. In addition, cost sharing of the planning costs has also limited the number of studies.

A separate cost sharing issue, different from the financing problems of coming up with the nonfederal share, is that of allocating the proportional share of project cost to the several purposes of a multipurpose project. The fact is that there are serious conceptual economic problems with actually implementing any cost sharing scheme for multipurpose projects. In economics, this is called the "joint product" problem: "In the case of cost allocations where products are jointly produced by the same facilities, no conclusive results appear to be possible, even in principle" (Herfindahl and Kneese, 1974, p. 290). Faced with this dilemma, different authorities have chosen to use different allocation algorithms. When applied to specific cases, however, these result in radically different cost shares. The most widely accepted method is known as "separable cost—remaining benefit" (SCRB). Under this method the strictly separable costs for each purpose are determined and then charged to that purpose; the remaining costs are allocated in proportion to the cost of supplying that purpose with the best single-purpose project. In this way some attempt is made to reflect the marginal value of that particular purpose. James and Lee (1971, p. 533) compare and contrast 18 different cost sharing methods.

The inability to find a unique "correct" cost share for multipurpose projects is likely to haunt the agencies in the near future as they attempt to implement the directives of the 1986 Water Resources Development Act. The prospect of different agencies using different rules for different parts of a large multipurpose project is likely to come about unless the federal government agrees upon a uniform set of rules and is determined to stick by them in the face of political challenges by local interest groups who stand to benefit from a different set of equally plausible rules.

Economics Applied to Water Quality Management

The externalities implied in water pollution were first extensively discussed by the Resources for the Future group in various publications in the early 1960s. By the early 1970s water quality had become a major political issue in the United States; with the passage of the Clean Water Act in 1972 it became a growth industry.

The public goods nature of water and the pervasiveness of externalities in water use lead inevitably to an increase of pollution as use increases, and by the late 1960s and 1970s, the public perception was that the environment was being overpolluted. It has never been shown what level of pollution is "overpollution," but the political perceptions were such that huge sums of money were appropriated for massive cleanups. However, since it is not possible to subsist in the environment without causing some damage (change) or without depleting natural resources, the question for society would appear to be not whether to pollute, but how much. The question has the earmarks of a political problem in social choice, yet economists have provided cogent analyses that help address the question.

The legitimate concern of economists analyzing water pollution is with market failure. In particular, the issue is how to internalize the externalities. It would seem that if the effluent could be correctly priced, then industry and other polluters could be taxed by just this amount and the problem would disappear—and the result would be the "correct" amount of pollution. Accordingly, the price could be raised until the satisfactory level was achieved. But as H. L. Mencken once remarked, "for every problem economists have a solution—simple, neat, and wrong." Given the simplicity of the effluent taxing suggestion there must be something seriously wrong with it since, where it has been applied in the United States, water pollution problems still remain. A distinction needs to be made, however, between sewer effluent pricing, which is commonplace, and taxing effluents that discharge directly into the environment, which is very rare in the United States and which has wider application in Europe (particularly Germany). The former is quite successful in inducing industries to reduce their water pollution loads, and the experience with the latter is that, in general, the governments are reluctant to raise the effluent charges high enough to achieve the desired levels of waste reduction. Moreover, since the experience also indicates that there is still a need for strong regulatory action on the part of government, the current approach seems to favor tradable effluent permits. Under this scheme industries are given permits for set amounts of effluent; they may discharge their effluent to the permitted amount, or they may choose to treat more of their wastes and sell the unused permits to other industries that may not be quite as efficient in removing pollutants from their waste streams. The one experiment with tradable wastewater discharge permits on the Fox River in Wisconsin (Downing and Sessions, 1985) produced inconclusive results. Nevertheless the 1991 Clean Air Act has chosen to use this technique for control of emissions of sulphur oxides by the electric power industry.

A simple example demonstrates the range of control options available for water quality management. Consider two industries on a river, with industry A upstream of industry B. A takes water from the river, uses it in its processes, contaminates it, and returns it to the river. B now has to treat the water before it can use it for its own processes. A therefore causes additional costs to B by its action of polluting the river. B has recourse to several approaches to redress this situation. The first and most obvious is to request A to cease polluting the river. If A agrees, the problem is immediately resolved—by either treating its wastes or changing its production processes, A can make sure that B is not damaged by its actions. A has thereby internalized the externality. What if A refuses? A might agree to negotiate with B so that both would share the cost of cleaning up the waste. B might pay A for treating the wastes or A might pay B to clean up water it needs for its processes. Depending upon the costs, the latter alternative could be a lot cheaper than having A treat all of the wastes. In this case the externality is internalized by building a theoretical "bubble" over the two plants and considering them as one.

From these types of examples, it is clear that externalities do not necessarily lead to economically inefficient solutions, provided that the polluters and those affected by the pollution agree to negotiate (Coase, 1960). It works when there is a consensus that the responsibility for damages is a reciprocal one, with the affected party taking steps to avoid them as much as the perpetrator to avoid producing them. Economic efficiency is achieved by having the costs of avoiding external damages borne by the party that can most cheaply end them. This solution, however, skirts the fundamental issues of equity and distribution. Negotiation may work quite well in the hypothetical case between two industrial plants of similar size, but it is unlikely to work as easily between one large industry and hundreds of thousands of private citizens. Both the asymmetry of the power to negotiate and the transaction costs of the negotiations themselves argue against achieving such a Pareto optimal solution.

Moreover, the idea of fairness, which seems to be missing in economists' ideas about equity, raises moral imperatives that are hard to avoid in dealing with externalities in practical cases. Take, for example, the problem of acid rain between Canada and the United States. From an ethical point of view, it is hard to argue that the damages imposed upon Canada by the United States are reciprocal; rather, the United States should be considered strictly liable for the consequences. The issue is how the United States should meet its liabilities: by treatment at source in the United States, or by paying the Canadians for the damages imposed, or by some combination of those strategies? Fairness as viewed by a noneconomist would suggest that the United States attempt to arrive at an economically efficient solution via least cost methods only after acknowledging its liability.

Returning to the two plants on the same river, what if plant A refuses to negotiate with plant B? At this point the process has to involve a third party. In the United States this would typically mean a lawsuit brought by B against A. A typical solution would have A forced to treat its wastes up to the current level of best available technology (BAT) for that particular industry. But B may still be forced to treat its intake water because the residual wastes now legally discharged by A have deteriorated its water supply. From this it becomes evident that the total costs may be substantially higher than the economically efficient solution. It is also apparent that the water pollution legislation and laws have created property rights for A to pollute the river up to a certain level.[1]

The current situation in the United States with regard to water quality is almost exactly the situation described in the last paragraph, except it is more likely to be citizens' groups suing the government to enforce the existing regulations against particular industries and, increasingly, against municipalities. This is certainly a long way from being economically efficient, but it is tinged with equity in that each industry type and each municipality is more or less forced to use the same type of treatment and, hence, face similar costs. However, because of economies of scale, small communities may face disproportionate per capita costs.

The hypothetical two-plant example can be helpful to demonstrate the principles. Figure 6.5 shows the costs for plant A to treat its waste to various levels of purity: 100 percent means that the water is returned to its original condition, and 0 percent means no treatment of the wastes. The figure also shows damages caused to B by A's action, measured in terms of the cost to B to treat the polluted water. The sum of these two sets of costs is also plotted and gives the total "social costs" of the pollution problem.

If there is no negotiated settlement to the conflict between A and B and the issue is resolved by regulatory action of the state, what level of treatment of A's waste should be chosen or mandated? Figure 6.5 shows the costs to each and also the total cost (to society if these are the only two members). The total cost curve shows a minimum at about 58 percent removal of A's wastes. This is the point where the marginal cost and the marginal damages curves cross; in economic parlance, where marginal costs equal marginal damages. This point, labeled O on the axis, is the point of maximum economic efficiency for this problem. At point N, the costs to A equal the damages to B. The gains and losses associated with the various solutions to this problem are as follows:

- At level P, A spends $0 and B spends $800,000 per year.
- At level M, A spends $800,000 and B spends $0.
- At level N, A spends $280,000 and B spends $280,000.
- At level O, A spends $180,000 and B spends $320,000.

Level M is the strict liability solution with A paying for the cleanup; level P is a laissez-faire solution with B absorbing all of the

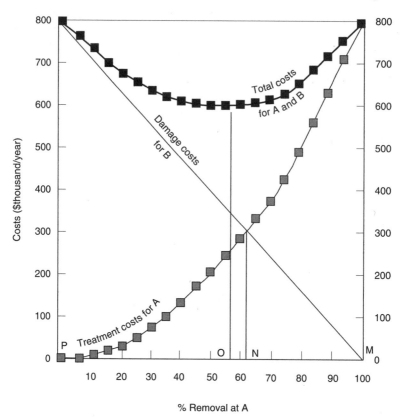

Figure 6.5
Treatment costs versus removal with two industries: the treatment cost curve for
industry A and the damage curve to downstream industry B as a function of the
percent removal of pollution at A. The top curve shows the sum of these two
costs.

damages; level N is the "equitable" solution with both sides bearing
the losses equally; and level O is the economically efficient solution.

Since it is unlikely that the negotiated solutions between the polluters and the impacted will come about spontaneously, some form
of government regulation is inevitable. The original economic dic-
tum of getting the prices right in the case of regulated externalities
now means setting a price upon the effluents of A such that the efficient solution, point O, is arrived at. What tax could be levied upon
A that would lead to this solution? If A produces 100,000 lbs of pollutant per year, treatment at 58 percent removal would leave 42 per-

cent, or 42,000 lbs in the effluent stream. At this level of pollution B suffers $320,000 of damages, so if A were charged $320,000/42,000, or $7.60 per pound of effluent, B could be fully indemnified for its damages. At this level A would treat its waste, because at 58 percent removal the marginal cost of treatment is $7.60 per pound. Below that it would be cheaper for A to treat and above that it would be cheaper to pay the tax. Hence, an optimal tax exists and the problem is solved.

Or is it? If it were not practical to reimburse B, B would agitate to move at least to the "equitable" solution at N, which is no longer socially optimal but at which B suffers less damages and the costs to each industry are the same. In a typical real-life situation there are a few large polluters and many thousands who are impacted. It is unrealistic either to expect them to negotiate with each other or to set an optimal effluent tax and then repay each individual according to his or her respective damages. The upshot of the measurement difficulties is that effluent charges have not been widely used in the United States for water pollution control. Instead effluent levels have been set based upon the expected performance of specific treatment technologies: the current requirement would force an "equitable" solution tilted in favor of the downstream user.

Empirical Studies of Water Demand

The concept of demand is clearly very important in estimating future investments in water. Demand studies from 1963 onward show price elasticities ranging from a low of −0.05 to a high of −1.38. The elasticities vary widely from region to region and from season to season, with low values and low prices in the moist eastern parts of the country and high values for summer water use in areas with higher prices. Short-run elasticities are more inelastic than long-run because the adjustment to conserving water often requires planning and capital investment on the part of the consumers (low-flush toilets and improved lawn watering equipment), and planners should not be led to think that the demand for water is therefore very price inelastic.[2]

Practical Pricing Problems

Pricing has a twofold role in water policy. First, increasing prices tends to ration water by cutting uneconomical consumption. This cuts the demand by moving up the demand curve, which is the effect most decision makers are looking for when pricing policy is advocated. The second aspect, and the one most frequently overlooked, is that of increasing the supply. When the price is higher, supplies of water from more expensive sources become available.[3]

The canon of price theory is marginal cost pricing, or setting the price at that point where the demand and supply curves intersect. Unfortunately, for investments that exhibit returns to scale, as most water infrastructure projects do, there are at least five different ways of measuring marginal cost, and each gives a different estimate. An excellent summary of these approaches is given in Meier (1983). Of the five major approaches, the simple marginal cost (based strictly upon the cost of supplying the next unit of water) can be eliminated right away because it would add the costs of increments of excess capacity to the next unit supplied, which is totally unrealistic. However, the remaining four each have strong arguments in their favor.

The problem with marginal cost pricing in the presence of returns to scale is best demonstrated by the intersection of typical demand and supply curves. The supply curve is a smoothed version of the stepped function of figure 6.3. The problem with a supply curve based upon figure 6.3 is that once the demand grows sufficiently to demand water from interbasin diversions, something quite odd happens to the supply. Local surface water diversions and groundwater supplies can be assumed to be almost continuously available from small to larger amounts with larger amounts costing more than smaller amounts (assuming away some of the lumpiness in pump sizes and diversion weirs), but once large engineering projects are required for interbasin diversion there are large fixed costs and the first few gallons of water supplied are enormously expensive. With the increase in demand, however, succeeding amounts of water become cheaper, with a marginal cost equal only to the prorated operation and maintenance costs of the project until the capacity of the project is used up. Then the cycle repeats itself; the next unit of water is very expensive, requiring huge capital costs, and subsequent amounts of water supplied require very little.

Regulated monopoly pricing. Even if the marginal costs could be unambiguously established, the pricing problem for most water uses is still not resolved because of the characteristics of water that make its supply a natural monopoly. The existence of large economies of scale ensures that the first entrant into the market will always be able to underprice and drive out any potential newcomers. It could decide upon the desired profit level and set the prices accordingly. This has been recognized for a long time in the area of municipal water and wastewater treatment and the suppliers have been regulated, typically by public utility commissions. Unfortunately most of these regulatory commissions tend to take a "backward" accounting stance and allow pricing based only upon average costs and the revenue needed to meet them. This is not appropriate in situations where the utilities are facing increasing marginal costs, that is, where all the best projects have been built and it becomes increasingly difficult to supply the same amounts of water for their historical cost. Under these conditions a "forward"-looking accounting stance is indicated. If this were pursued then the revenue requirements of utilities would focus on establishing adequate future investment funds.

In the event, however, another problem would surface, namely that the utilities would raise substantially more revenue than needed to cover their operating costs and retire historical debt. Serious administrative and legal problems could arise because of this.

It has been suggested that water be priced to motivate rational consumption during the year, and that at the end of the year customers receive a rebate from the water utility returning any surplus funds. Alternatively the surpluses could be put directly into general revenue accounts of the municipality or placed in separate environmental enhancement funds. Whichever approach is taken, it is important to note that these funds should be an integral part of any recommendations for a move toward marginal cost pricing.

Municipal water pricing. A decade ago, Mann observed: "Traditional rate-setting methods, employed by state regulatory commissions as well as local government agencies, appear to have produced a situation of rapidly deteriorating water systems, both rural and urban, characterized by aging capital facilities and under-maintained water systems" (Mann, 1981, p. 101). Since then the tightening of the federal drinking water standards and the reduction of federal subsidies to wastewater treatment have made the situation even more pressing. The American Water Works Association's manual (1972) on municipal water pricing reflects traditional rate-setting procedures of recovering the costs of providing water by use of multipart tariffs based on average cost pricing. This approach results in an overestimation of future demand and overbuilding.

The definitive work on municipal water pricing is the excellent study of Tucson, Arizona, by Martin, Ingram, and Griffin (1984), which chronicles the role of water in the life of a desert city. (The Tucson city government was recalled by the electorate because of an increase in the price of water, probably the first time this issue has led to the downfall of a city government anywhere in the world.) The authors analyzed the political, social, economic, and technical issues underlying water and its use in Tucson and presented the generic issues of water planning for city managers elsewhere. Despite the highly political nature of decision making about water, and water pricing in particular, they acknowledged the potential usefulness of economic reasoning. The reason economists have not been more successful, they wrote, is that "they have not provided sufficient concrete examples of how to apply their economic theories, nor have they spoken a language or through channels to which water managers are receptive."

Effluent pricing.   Regulating environmental quality by pricing the effluents that individuals, municipalities, and corporations emit is one of the great debates that never took place in the United States. The economic literature is full of conceptual schemes for pricing or taxing effluents and selling pollution permits that would lead to internalizing the externalities of pollution. Once this is done, environmental quality can be left entirely to the usual market forces. The Washington-based research institute Resources for the Future has led the campaign for fees on effluent discharge for over 20 years with little success, opposed primarily by those who think it improper to sell the right to pollute the environment.

In fact, both positions are unrealistic; both ignore some basic imperatives of the environment. Even if it were possible to figure out the optimal prices or taxes to impose, we would still have to monitor the actual performance of the polluters to make sure they were not cheating. Hence, we would still need large and intrusive bureaucracies to implement the market solution. Under these conditions, how efficient would the "market" be? On the other hand, the mainline environmentalist position does not recognize that under the current regulatory system, permits to pollute are in effect issued unless no emissions at all are permitted.

Despite the lack of debate on the issue in the United States, a significant amount of effluent pricing has been carried out here for many years in the municipal wastewater sector. One report claims that by 1970, more than 90 percent of the municipalities with populations over 50,000 levied some form of sewer charge on residences and industry and that 40 percent of local expenditures on sewage were derived from these charges. (Residential charges are not relevant here since once hooked up to a sewer, residential users have few options to adjust their effluent quantities. The sewer charge is typically set as a fixed proportion of the water bill; the only option is to reduce consumption.)

Industry can respond to effluent pricing by pretreating its waste, reducing its water use, improving its housekeeping, changing either its production processes or products, or by paying the fee. The industrial charges are typically related to the quantity of water used and the strength of the effluent, typically measured in terms of the oxygen-demanding organic waste load and total suspended solids. These charges can give industries an incentive to change the amount and strength of their sewage effluents. A study by Hudson (1981) of 5 cities (Atlanta, Chicago, Dallas, Salem [Oregon], and South San Francisco) and 101 industries found that effluent charges were overwhelmingly preferred to discharge limitations by the industrialists.

The responsiveness of industry to increased effluent charges is dramatically demonstrated by the 70 percent reduction of process wastewater by the Clairol, Inc., plant in Camarillo, California. Woodman (1989) reported that Clairol's response was stimulated by the closing of a local land disposal site that raised the cost of waste disposal from $0.25 to $1.50 per gallon. In the subsequent investigations by the plant engineers it was discovered that, when the value of discarded product and the cost of purchasing the water in the first place was added to the new off-site disposal cost, each gallon of waste cost Clairol $33. This surprise led to a major rethinking of the wastewater management system in the plant. Most of the water-carried systems were replaced with air cleaning systems which saved water and allowed for easy recovery of chemical feedstocks. One of the systems that cost $50,000 to install is saving the company $240,000 per year. This example is typical of what might be expected in many other industries.

Development of Private Water Markets

Water ownership rights in the United States are in a state of chaos. Much of the discussion about rational planning and allocation of water, and about water quality, has focused upon these rights. The conclusion is often that if the resources could be privatized, the problems would resolve themselves. Recent experience with the California Drought Bank strongly supports this argument. The discussion of marketing water rights takes place against a background of conflicting trends in property rights in different geographical regions of the country. A study by Saliba and Bush (1987) summarizes recent moves in this direction in the southwestern parts of the United States.

Markets for water develop when potential buyers and sellers perceive that there are economic gains to be made by transferring ownership of the resource. In mature commodity markets the magnitudes of these gains are very small, being restricted to the returns on capital and management. (When markets have been in operation for long periods of time both buyers and sellers know exactly how much each can benefit by the transactions.) There are usually few new applications for which resource use is highly constraining, hence the potential for scarcity rents is small. For water, however, where there have been many different "markets" with radically differing prices for the same commodity, there exists the possibility for large profits to both buyers and sellers.

Farmers who receive federally subsidized irrigation pay only a fraction of what municipalities and industries pay, or are willing to pay, in the same market area. In some cases a farmer pays as little as $3 per acre-foot for raw water while the municipality pays more than $300. Clearly, there might be some benefit to both the farmers and the municipalities by trading some of this water. This makes even more sense when it is understood that the value of irrigation water is typically less than $50 per acre-foot in terms of its marginal contribution to agriculture, or very small compared with the economic returns to water use for municipalities or industry. (This shows clearly on the demand curve of figure 6.2.)

In addition to the existence of these disparities in cost and value there has also been a rapid increase in the urban populations, with a commensurate increase in the demand for services, notably for water supply. These differences in value have been exploited during the recent drought in California to reallocate water from low-value to high-value uses.

The character of the water market is changing rapidly in the West, particularly in the Southwest, where the demand is for urban water uses. But the system lags behind the realities: in 1980, irrigation consumed 89 percent of Arizona's water, mines less than 3 percent, and all the other uses 8 percent. Agriculture, however, contributed only 2 percent to Arizona's personal income, with the remaining 98 percent coming from those sectors that consumed less than 11 percent of the water (Welsh, 1985). According to this, a transfer of only 5 percent of the water currently used in Arizona's agriculture could support a population increase of 50 percent (another 1.5 million people). The same sort of conclusions apply to each of the so-called water-short western states.

Several observations can be made about the scarcity of water in the Southwest by region and by time period. First, of the five states examined by Saliba and Bush, Arizona appears to have by far the cheapest water (least scarce). There appear to be large spatial and temporal differences between the different transactions as well, although the highest water prices in several of the regions were reached during 1980 and 1981 when there was widespread uncertainty with respect to natural resources because of rapidly increasing oil prices and other inflationary pressures. More recent prices are substantially lower than those of 1980–1981.

Water is available in private markets in the Southwest valued at between $100 and $200 per acre-foot on an annual basis. These figures are lower than might be expected considering the cost of developing new water supplies for cities, but are still substantially above the marginal value of water in agriculture. The potential, therefore, for continued water marketing in the Southwest appears good.[4]

Why Bother with Economics?

This chapter could be read as an assertion of the futility of attempting economic analysis on a subject as complicated as water policy. We have great difficulty in measuring the benefits and disbenefits of development actions: indeed, if you believe in the reality of the intrinsic benefits then typically 50 percent of the benefits are completely ignored. Marginal cost pricing is not strictly possible and the plausible alternate methods for approximation give radically different policy implications. Cost sharing for multiple-purpose projects is not even possible in principle, and again plausible pragmatic alternatives can give different answers. Finally, pricing of pollution externalities appears to be theoretically possible but is an administrative nightmare that is rarely put into practice.

These problems appear to make economic approaches to water policy irrelevant. Why then has there been so much attention focused precisely upon the economic aspects of water? The answer is simply that economic thinking and conceptualization are the only alternatives to a chaotic political battle with no clear concept of the public good, only logrolling and pork barreling.

Moreover, some very important economic phenomena are at work. First, the market seems to work quite well in allocating scarce water, specifically in the West. In fact, it works better than most economists themselves would have predicted only 10 years ago. Second, water consumption is clearly price responsive. The problem is to find some reasonable second- or third-best pricing schemes (Hanke and Davis, 1973). Finally, the creation and the continued existence of revolving state funds for water quality indicate that there are mechanisms for tapping a very real willingness to pay for water quality.

In sum, while economic analysis and economic thinking by no means solve all of the problems in the field, water managers and consumers must apply them if a coherent water policy is to emerge in the United States.

7

Political Imperatives: Legislative, Executive, and Bureaucratic

Economic imperatives establish the total cost of a water investment or policy to an entire society or to subgroups within society; political imperatives stand between the somewhat abstract economic analysis and actual realization of government action. Political imperatives contrast to the economic imperatives in three important ways:

1. They concretely evaluate the desirability of a policy or an investment on the basis of its value, positive or negative, to a large number of subgroups with varying degrees of interest in the matter.
2. They do not rely solely upon the simplifying quantitative economic measure of money, but are heavily influenced by nonmonetizable considerations as well.
3. They are pursued separately from economic objectives, with different personnel and rituals; recruitment to the political arena has a particular history and admission confers a great deal of authority.

At the federal level, the traditionally dominant channel for political decision making in the water domain is the U.S. Congress. In a fascinating and highly personalized continuous process, it apportions weight to the interests of different groups (as well as to technical changes, economic factors, and intersectoral relationships) and periodically announces changes in national goals.

The Institutional, Legislative, Political, and Legal Imperatives of Water Policy

The institutional, legislative, legal, and political imperatives are difficult to specify because they typically fuse many issues at once. Their collective contribution, however, is to impose a socialization process on the technical, economic, and financial imperatives.

The social and political imperatives start with equity, and with those market failure issues that elude economic and financial approaches. For example, political processes must defend the principle that no group or individual in society should bear a disproportionate cost in obtaining the basic water required for survival or for meeting environmental requirements; they must also work toward the elimination of wasteful or inappropriate use of public subsidies.

Society must guard against irreversible environmental degradation which may come about because of the divergence between the present economic values of resources, placed upon them by the current generation of citizens, and their value to future generations. For example, due to pollution in a lake the ecosystem may be so radically changed that it would become eutrophic, choked with vegetation, and eventually die. Even if the current generation does not think it worth saving, it falls to the body politic to take action to preserve the lake so that future generations are allowed to make that decision. As a society we are ambivalent in setting such values on environmental amenities in the same way.

Political mechanisms have to be developed to guard against the technological imperatives of scale and location. Technocrats cannot be permitted to run roughshod over minority or local interests and build big projects in the "best" locations, such as the Grand Canyon. Economists argue that option values would avert these risks, but until such time as the practical application of option values is much more widely used and understood by the public at large, there is a pressing need for political fixes.

Two aspects of political management seem particularly obvious: the need for clear and easily implemented legislation and for active consent to the legislation by the citizenry. Unfortunately, because of the complexity of the interests involved with water, it is hard to write such legislation. And once regulations become as complex as they are now, active consent and participation become difficult to obtain. Without voluntary compliance the regulations become more difficult to implement and lead to more complex regulatory actions, essentially ad infinitum.

The politics of water are invariably more intense at the state and local levels than at the federal level. At the local level, who benefits and who loses as a result of water policies is painfully obvious. Many local political issues, however, are fought out at the federal level, often in transmuted form, because of the pervasive role of the federal government in water policies. President Carter's 1977 list of western water projects to be canceled was fought by the western senators in Washington even though the real issues dealt with the allocation of water between agriculture and municipalities within the individual states. Localities attempting to resolve very local political issues of water quality appeal to the federal government for relief: cities like Boston, which have successfully avoided meeting the federal water quality laws for over 15 years, are now trying to blame the federal government for the very high cost of finally bringing their systems into compliance with the law.[1] Boston has yet to have a serious debate about who is benefiting and who is paying for water quality improvements. In this and many similar cases, the local political process attempts to use the federal arena to avoid dealing with unpleasant subjects.

Because of the new emphasis on non-point pollution, local upstream-downstream conflicts will likely become much more pronounced in the near future. These can only be resolved at the local level within, however, a federal framework. Hence the need for the clarification and improvement of federal water policy.

Institutions. The long-term maintenance of societal institutions is also an important imperative. There must be agencies and other organs of government to carry out regulations. They must be efficient, meaning that the cost of implementation must be small in relation to the size of the overall problem. There must be clearly understood penalties for noncompliance and a mechanism to enforce the law equitably and swiftly. Cooperation between environmental agencies and other governmental departments needs to be practiced and routine, that is to say, institutionalized. But institutions also look after their own survival, often outliving their utility or the goals that brought them into existence. It is not realistic, therefore, to assume that institutions or bureaucracies exist only for the common good.

Physical imperatives may demand new legislation or major changes in the laws. Physical characteristics, such as the (very inflexible) hydrological and nutrient cycles, and technical possibilities in the field of metropolitan water supply and wastewater disposal, for example, should be given greater attention. We need to get away from overly restrictive rulemaking that forces the adoption of one or two specific technologies, as is currently the case with municipal sewage treatment plants.

Legislation.   Much confusion could be alleviated by a conscious attempt to analyze the diverse imperatives that enter into policy decisions over the natural systems. Laws, institutions, and economic incentives that get in the way of effective functioning of these systems may have to be modified—for example, supporting private water markets in the irrigated areas of the arid West. Private marketing is already happening; all that is needed is the political commitment to adjust the laws and the institutions to allow the free sale of publicly and privately supplied surface and groundwater, to be treated equally.

Models of the Federal Process

Various authors have suggested models of how the federal government works, albeit very loose ones by the standards of the physical and economic sciences. Pastor (1980) characterized the four major approaches as follows:

The bureaucratic politics and process model.   This model is based on political-bureaucratic bargaining. Its focus is typically the executive branch, with Congress hardly in the picture. Classic cases are drawn from foreign policy problems (for example, the Cuban missile crisis) where Congress was not a major player, but this is the opposite from the situation in water, where the executive branch until recently was largely excluded by Congress (Maass, 1951).

Although there is a long history of conflict between the Corps of Engineers and other executive agencies, particularly the Bureau of Reclamation and the Department of Agriculture, there is no reason to believe that this model will give the levels of description, explanation, and prediction that others provide.

The congressional behavior model.   A second model concentrates on Congress, with the view that to understand congressional behavior is to understand that congressmen are "single-minded seekers of reelection" (Porter, 1980, p. 36). It follows from this that congressmen's goals are to improve the welfare of their constituents in the shortest possible time frame. The surprise is that Congress as a whole is much more coherent and responsible toward national policy than such individual behavior would suggest.

Another major determinant of congressional behavior is the committee structure, in particular the operating methods of the Ways and Means, Appropriations, and Rules committees. This aspect is indeed very important in the field of water policy.

Twenty-seven committees and more than 184 subcommittees have some jurisdiction over water issues. Overlapping control among committees confuses and delays the passage of legislation and leads to inordinately repetitious hearings, the same congressmen often hearing the same witnesses present the same material to two or three different committees. Analysts have discerned three emphases in committee behavior: those (like Ways and Means) whose members are principally "insiders," oriented to influence in the House and legislative success; those (like Interior and Post Office) whose members are constituent-oriented; and those (like Education) whose members are primarily interested in public policy. In comparison to the House, Senate committees are less important as a source of chamber influence. For water policy, the House Interior and Insular Affairs Committee and its subcommittees, and the Senate Environment and Public Works Committee and its subcommittees, are the most important forums for constituent and interest group activities.

The realities of information processing are also important in describing congressional behavior. With humanly limited capacities to absorb and judge, legislators are so overloaded with information that they have to be extremely selective in committing their attention. Congressmen deal with this by specializing in a particular and limited area; in other domains they take their cues from other sources (colleagues, outside groups, committee reports) that they have learned to trust.

The interest group model.   When a national legislator thinks about the constituency that elected him, he rarely if ever sees an undifferentiated mass of individual voters. He sees categories of interests. In some cases, he sees only a few dominant interests.

There are literally hundreds of active interest groups—environmental groups, water resources groups, professional associations, and industry associations—involved with water policy. These interest groups often have overlapping concerns and overlapping memberships. They constitute vital channels for particular publics to participate in the federal governmental process. Pork barrel projects are the fodder for the well-known "iron triangles" of legislators, bureaucrats, and active interest groups that develop in specific issue fields (Heclo, 1977). (The term "pork barrel" was first used to describe the exchange of benefits in successive Rivers and Harbors acts, usually referred to as omnibus bills.)

In the water area, an old idea of the "concurrent majority" (Calhoun, 1853) may be much more relevant than newer views on the interest group model. Under this older concept it was recognized that major government policy decisions must be made with the approval of the dominant interest groups directly affected. It appears, for example, on federal water quality legislation, that the Sierra Club, the Conservation Foundation, and the Environmental Defense Fund are actively encouraged to give their approval before legislation can be passed.

The interbranch politics model. The first model dealt with bureaucratic politics, the second with congressional behavior, the third with the interactions of Congress, bureaucrats, and interest groups; the only important area not yet covered is that of the interaction between the executive and Congress. Arthur Maass (Maass et al., 1962), who has written extensively on water policy issues, stresses the relationship among the constitutional branches of the federal government. Powers swing back and forth between the branches depending upon the particular issue.

Maass sees the issues as being between two institutions, Congress and the executive; between two processes, legislative and administrative; and between two roles, oversight and initiation. His general rule is that the executive plays an initiating role in both the legislative and administrative processes and that Congress oversees both processes. Other observers, influenced by the bureaucratic politics model, would argue for considering the bureaucracy as an independent political entity. This would broaden interbranch politics to three "branches" rather than two.

A Case Study of How Well the Models of the Federal Process Predict Water Policy

It is not necessary to consider these political models as mutually exclusive alternatives; it is clear that all contribute to the outcomes of federal policymaking through a process of constant competition and synthesis. The usefulness of the models in predicting outcomes, however, is difficult to determine, since it is rare to find a piece of federal policymaking that is sufficiently discrete as to be clearly identifiable throughout its development. In water policy, however, there is such a case, brilliantly isolated and described by T. R. Reid (1980). The particular piece of legislation, though relatively minor, did have some important constitutional ramifications; but above all it has been presented in complete, transparent detail. It is, therefore, an excellent case upon which to reflect how well the models sketched out above describe actual political behaviors.

Reid gives a blow-by-blow chronology of the events as they unfolded, documents the complexities of the action, and plumbs the motives of the actors. Reid was assigned in January 1977 by the editors of the *Washington Post* to "choose some little-known but important bill, and follow it week by week as it moved along toward passage or defeat" (Reid, 1980, p. ix). The bill eventually chosen, S.790, concerned user charges and capital recovery charges for the federal inland waterways system and authorization of Lock and Dam 26 on the Mississippi River, a $400 million project on the Mississippi River at Alton, Illinois. It was filed by Senator Pete V. Domenici (R, New Mexico) on February 24, 1977.

Like many congressional bills, this one included more than meets the eye. It contained a number of seemingly disparate actions: Domenici wanted to set use charges for the federal inland waterways system, which the system's users did not want, but they did want to have Lock and Dam 26 built. Moreover, the issue was not confined to simple economics; environmental groups had long been concerned with the damming up of free-flowing rivers and its consequences for wetlands and fauna and flora along the rivers. They were, therefore, strongly against any federal aid to build a dam.

Reid documents how the concept of free and untaxed use of inland waterways goes back to the very founding of the Republic. The Northwest Ordinance, passed by Congress in 1787, included the provision that inland waterways would be "common highways and forever free . . . without any tax, impost, or duty therefor." "Forever" lasted from 1787 to 1938, when President Franklin Roosevelt ordered a study on taxing waterborne commerce. Roosevelt's study was lost in the greater concerns of World War II, but the idea survived. Each succeeding president asked for, but did not receive, a waterway user charge, the last request being a belated one sent on January 19, 1977, the day before Carter's inauguration, by the outgoing Ford administration.

Filing bills for waterway user charges had become one of the perennial, and seemingly doomed, activities in both houses of Congress. Domenici's strategy was brilliant. As a freshman senator, Domenici had been assigned to the Senate Environment and Public Works Committee and its Water Resources Subcommittee. There he was introduced to the workings of federal water resources policy. In particular, he was tutored by one of the committee staff, Hal Brayman, on the free ride that barge operators were getting at the expense of the taxpayers. Brayman had been arguing for years against spending federal taxes for an industry that could now well afford to pay its own way. Domenici was a perfect ally for him because New Mexico is a state with plenty of railroads and no barge lines.

Domenici saw his opportunity when a $400 million Corps of Engineers project to rebuild Lock and Dam 26 on the Mississippi in Illinois came up for hearings in the Subcommittee on Water Resources. These structures were in bad shape, causing delays for the shippers, who were eager to have them rebuilt. Domenici's bill thus coupled something that the barge interests desperately wanted with the last thing they wanted—a user charge. It was a classic "hostage bill": if you want the good thing you have to swallow the bad.

The barge shippers constituted a formidable interest group. Its congressional supporters included some of the most powerful committee chairmen in the Senate and the House.[2] In fact, the barge industry never seriously worried about the perennial user charge bills.

The chronology of the events from first filing on February 24, 1977, to the final passage on October 13, 1978, is given in appendix 6.

Bureaucratic politics.   The role of bureaucratic politics in this legislation centers on two people. The first was President Jimmy Carter, who as governor of Georgia had run up against a Corps of Engineers plan to dam one of the state's most scenic rivers. He had great difficulty in stopping that particular plan and, as a result, he harbored an unresolved animosity to the Corps that later led him into serious political blunders. Carter was in favor of the user charge but against the construction of the lock and dam, which the White House thought needed "further study." The second person was Brock Adams, the new Secretary of Transportation and a major player in getting this legislation passed. Adams, a former six-term congressman from Seattle, was an outsider in the Carter White House who owed his job to the urgings of a congressional group, led by Speaker Tip O'Neill, rather than of the White House staff. He was a longtime skeptic about the barge lines, but the political reality of reelection from a barge state had kept him from pursuing the issue. As Secretary of Transportation he finally felt able to move on it.

There was a wide variety of responses from the different federal bureaucracies to the proposed user charge. The EPA thought the idea was excellent, but the Maritime Administration (MARAD) of Commerce was opposed. The Agriculture Department wavered between neutrality and opposition, influenced by the large volume of agricultural products shipped by barge. Predictably, the Corps of Engineers was against the legislation. The Federal Power Commission was worried about the impact on coal prices and the Department of Transportation worried about the problems of keeping the barge transport prices too low, with potentially deleterious impacts on rail and road transport.

The White House, through the Office of Management and Budget (OMB), assigned the departments of Transportation and the Army to work out a single administration position on the Domenici bill. Transportation Secretary Adams and Clifford Alexander, Secretary of the Army, agreed without much difficulty that the user charge was worthy of support but that the lock and dam did not need rebuilding in any great hurry. The rest of the bureaucracy went along, including Major General Ernest Graves, head of the Corps's Civil Works Directorate. While testifying in favor of the administration position Graves admitted that "I would be less than candid if I did not call the committee's attention to the fact that . . . I testified before the committee last year to go ahead" with the new lock and dam (Reid, 1980, p. 34).

Adams's most serious problem within the executive branch was OMB's insistence that the waterway fees be collected as a "fuel tax." Adams was well aware that the use of the word "tax" would send the bill into the unwelcoming arms of Senator Long's Finance Committee, but his objection was overridden by OMB because the president's general policy was to tax fuel products. This "fuel tax" phrasing finally helped kill the original version of S.790.

Another episode of bureaucratic jousting occurred in the spring of 1978. Adams had promised Domenici a strong threat of veto from the president if the revised bill (now H.R.8309) was passed without Domenici's capital recovery clauses. Because of Adams's weak political position within the executive, all he could come up with was a letter signed by the president's staff secretary indicating the possibility of a presidential veto. This was not good enough, and on May 3,

1978, the Stevenson-Domenici version of H.R.8309 was voted down by the Senate, which obviously did not take the veto threat seriously.

Congressional behavior model. Much of the discussion of congressional behavior focuses on the House of Representatives, whose members must seek reelection every two years. The 1977–1978 inland waterways fee struggle is out of this pattern in that it originated in the Senate, where the reelection pressure is less immediate but still significant, as the case shows.

The first aspect of the congressional behavior model involves improving the welfare of constituents by voting for their interests. The Senate minority leader at the time of Domenici's bill, Howard Baker from the important waterway state of Tennessee, gave the bill a big push when he announced in favor of it three days ahead of the June 22, 1977, vote, but could not bring himself to vote for the revised version of it a year later when he was up for reelection. Timing is crucial; not only did Domenici lose votes to other senatorial reelection campaigns, but also to the pressures brought by specific lobbying groups such as the farm lobby, which converted Senators Dole, McGovern, and Hansen to the opposition.

Facing a vote on legislation like this puts senators under cross pressure. They cannot vote without hurting one important group or another. Some states, like Illinois, are railway as well as waterway states, and railroads favored waterway fees since they would tend to equalize costs with their barge competitors. During the June 1977 vote Illinois Senator Charles Percy (R) tried to satisfy both sides by delaying until the last moment when it was clear that the Domenici bill would win, then voting against it. His vote supported the barge interests but he could explain to the railways that he had waited until he was sure that they were going to win before he cast his vote. This tricky maneuver apparently caused him so much trouble during his reelection campaign in 1978 that for the May 3, 1978, vote he thought it safer to side with his Democratic colleague, Adlai Stevenson III, and support the revised fuel tax bill. For the bulk of the senators, however, there was no overwhelming sense of their constituents being clearly better or worse off by their vote on this particular issue; hence the simple interest-linked voting model does not explain congressional behavior very well here. Other factors were present.

A second major determinant of congressional behavior is the work of the large, nonelected professional staff, employed by individual senators and representatives as well as by committees. If Domenici had not been assigned to the Public Works Committee he probably would never have become interested in such an arcane subject as waterway user fees. But without the professional knowledge and policy drive of his two staffers, Hal Brayman and Lou Rawls, Senator Domenici would never have initiated the user charge legislation.

Senator Domenici immediately ran into the third major determinant of congressional behavior, the standing committees of Congress. In this case, the assignment of the legislation by the parliamentarian to the appropriate committee could have led to its instant demise. If the user charge were to be considered a "tax," it would have to be sent to the Finance Committee chaired by Senator Russell Long of Louisiana, an invincible foe of waterway charges. His committee would never have reported it out to the Senate floor for debate. If the user charge were considered a "fee" on inland waterway transportation it would go to the Commerce Committee chaired by Warren G. Magnuson, also an opponent. In addition, the chairman of the appropriate subcommittee of Commerce was none other than Russell Long. However, Domenici was able by some very clever parliamentary moves to get his bill jointly assigned to his own committee, Environment and Public Works, as well as to the Commerce Committee. Under a joint assignment, it could be acted upon by the Senate if it were reported out of only one committee.

The most important effect of the structure of Congress in this particular case, however, occurred when Domenici's bill had already passed through the Senate committees and had been approved by the Senate by a vote of 71 to 20. The bill had then been routinely inserted in H.R.5885, a House omnibus rivers and harbors act that had been recently sent to the Senate. This enabled the senators to have their user charge acted upon directly by the House conferees, after which it would in all probability pass on the House floor. The opponents of the bill (the barge lobbyists), however, immediately raised the issue of the "origination clause" with Representative Al Ullman, the chairman of the House Ways and Means Committee.

The origination clause is based upon Article I, Section 7, of the U.S. Constitution, which states: "All Bills for raising Revenue shall originate in the House of Representatives." This clause had caused a fair amount of conflict between the houses and in the courts during the nineteenth century until the Supreme Court maintained the distinction between legislation that was mainly regulatory, with the revenue-raising aspects only "incidental," and legislation that was mainly revenue-raising. S.790/H.R.5885 was clearly in the former category; nevertheless, Ullman made a formal request that the legislation be held at the Speaker's desk and asked the Clerk of the House to prepare the "blue slip" by which the House traditionally notifies the Senate that the origination clause has been violated.

On learning this, Brock Adams approached his old friend Speaker O'Neill to arrange a compromise that would keep the legislation moving but assuage the House's sense of its constitutional prerogatives. The compromise arrived at was a masterpiece.

To avoid the origination problem, O'Neill said H.R.5885 would be left to die and the House would originate its own version of S.790. The House Public Works Committee would approve a new lock and dam authorization. Ullman's committee would write a waterway tax bill, setting the charge considerably lower than the Senate version. The two measures would be linked on the House floor, passed, and sent to the Senate as a new bill. It had taken the Senate five months to achieve that much, but O'Neill said he couldn't wait that long . . . the issue was to be reported out of both committees within two weeks. (Reid, 1980, p. 74)

What happened to Domenici's bill as a result of this sharp turn was that the concepts of user and capital recovery charges were quietly being replaced by a very low flat tax. The compromise worked and H.R.8309 was passed on October 13, 1977, and forwarded to the Senate, where it became the framework for Senate omnibus legislation and where it fell victim to another feature of congressional behavior.

This feature, which originated in the water resources area, is that of the pork barrel, or legislation in which the various members add their favorite projects together for one mammoth bill. It is considered bad form to object to other members including projects when you yourself are going to add one or two that focus on some particular group of constituents. In the Senate, H.R.8309 was soon covered with more than 90 amendments totaling more than $2 billion in water projects. At this point Senator Long was able to convince the Senate that the threatened presidential veto would not take place if the capital recovery feature of Domenici's bill were dropped. Long's amendment was passed on May 3, 1978, and sent to the House. Realizing that his own pork barrel bill H.R.5885 had died, O'Neill refused to appoint House conferees until the House's version was ready. When it was, at the end of June, a further $2 billion in water projects had been added. President Carter, who had been badly burned by his attempt to cancel water projects in 1977 and was still smarting, threatened to veto the whole bill. As a result of this, and to get as far away as possible from the $4 billion pork barrel, those on both sides of the user charge issue agreed to start with a whole new clean bill.

The new bill incorporated the idea of a flat tax that now had been firmly established by the House version, and by the Senate opponents of the bill, but replaced the capital recovery aspects with a trust fund. The trust fund idea was clever because it did not limit congressional action to the amounts in the trust fund, but the tacit understanding was that it would be hard to get Congress to appropriate more than that amount. So a de facto limit would be established *and* the barge interests would know those limits and thus make their demands for federal investments in proportion to the most efficient use of those funds—precisely the original notion of capital recovery.

There were several more twists and turns before the final adoption of the new clean bill. The most important of these was the ability of the chairman of the Finance Committee always to have at his disposal a tax bill that had been approved by the House but not yet by the Senate, onto which fast track legislation could be appended and sent back to the House. (The bill actually used for this purpose, H.R.8533, dealt with tax exemption for charitable bingo games.) This meant that under the extreme conditions of the adjournment of the House on October 14, 1978, and within only four days of getting agreement from a broad cross section of the interested parties, Long was able to pass the legislation in the Senate and forward it to the House. If this had not been hostage legislation it is highly unlikely that this could have happened.

Another last-minute twist was an attempt by Senator Adlai Stevenson III to make an end run around the proposal by slipping through an authorization of Lock and Dam 26 as an amendment to some other legislation. This move galvanized both Domenici and Long into a frenzy of action that led to the final passage of the user fee legislation.

The congressional behavior model, then, helps explain the *how* of this bill. It is not very instructive on the *why*. For this it is necessary to look at other models of the federal process.

Interest groups.   Many people see the interest group theory as the best guide to federal policy in the water area, where outcomes often reflect the relative access, economic resources, and organizational skills that different groups enjoy. This particular water policy exercise was a good example of the interaction between interests groups and the Congress.

Waterway shippers were fully alerted that after almost 200 years of guaranteed federal immunity from bearing any part of the now appreciable costs of maintaining and improving the waterways, they were faced with sudden, substantial, and permanent charges.

The National Committee on Lock and Dam 26 was formed by the barge interests to coordinate their legislative efforts. The National Waterways Conference, the Association for the Improvement of the Mississippi River, the Upper Mississippi Valley Association, major waterway shippers including oil, chemical firms, and farmer organizations, and shipyards and other service industries dealing with the inland waterways fleet constituted the major interests opposed to the bill. Their most important instrument against the user charges was the American Inland Waterways Committee set up by Louis B. Susman, a St. Louis lawyer with a reputation for hard bargaining, who was also a member of the Democratic National Committee. Susman recruited the Timmons Company, the chief lobbyists for the Nixon and Ford administrations, former Democratic Senator George Smathers from Florida who was until then a $100,000 per year lobbyist for the railroad interests, and former Democratic Congressman James Symington from St. Louis. In addition, there were congressmen and senators from states that had significant mileage of waterways, and tacit support from government agencies (the Corps of Engineers, the Maritime Administration, and the Federal Power Commission) that did not wholeheartedly support the Carter administration position.

The railways supported the legislation, since they were clearly eager to see the water shippers' cost advantage ended. They worked through the Association of American Railroads and the Western Railroads Association. Aligned with the railways were a variety of environmentalists and good government advocates who wanted the costs of investments to be borne by the beneficiaries, as a matter of principle.[3] The most important pro-legislation coalition was the Council for Sound Waterways Policy, set up by J. D. Feeny, the general counsel for the Western Railroads Association. Reid refers to this organization as "Feeny's Laundry," because it was set up to launder the contributions of the railways to groups that did not like to be seen taking railroad money. Reid claims that the Public Interest Economic

Center received a $35,000 grant, the Environmental Policy Center received $1,500 per month, and other groups received donations exceeding $5,000 per month from the railways to pursue their work on this issue.

In the final reckoning, each side spent at least $500,000 in presenting its case. Both made errors. Initially the railroad interests, for example, focused solely on trying to stop Lock and Dam 26 and ignored the user charge issue. The barge interests were so intransigent that they ended up alienating their major Senate supporter, Russell Long. Interest group analysis in this case witnessed a crowded and active scene, but appeared to have little predictive power.

Interbranch politics. Arthur Maass, one of the leading writers on the politics of federal water policy, favors the interbranch model of the federal process, which seeks to explain policy outcomes in terms of the effects of Congress and the executive on each other. Specifically, it asks what the policy priorities of each branch were in a given case, and how the conflicts were resolved.

In the waterways case, it is hard to assess what the specific congressional policy priorities were a priori, except that overall Congress wanted to continue the doctrine enunciated almost 200 years before in the Northwest Ordinance—forever free. On the other hand, there had been a growing awareness among some congressmen that equitable treatment to all taxpayers clearly demanded user charges.

In the executive, there appeared to be more conflict among its own agencies (notably the Corps of Engineers) than with Congress. This may have been due to the efforts of Secretary of Transportation Brock Adams, who took a major role in keeping the houses of Congress working together. Without his interventions, the conflict on the origination clause could have resulted in a major blow-up between the houses and the bill certainly would have died. But in this case, the executive's ability to insert itself between the two houses was entirely due to the credibility of Adams, based on his six terms in Congress.

The outcome of the bill certainly was the adoption of the low rate of tax and a decision to proceed with the construction of Lock and Dam 26. (By 1992 the construction of the lock was well under way at a federal cost in excess of $400 million.) This was some, but not all, of what the executive, Senator Domenici, and Hal Brayman had wanted. The limited scope of victory for the White House was due in large measure to the fact that President Carter had severely compromised himself by the injudicious way in which he had fought a set of large water projects the year before, and indeed made a frontal attack on the very concept of pork barrel legislation. This made the Congress wary of the new administration, and when Carter backed down, his credibility in Congress was severely damaged. Hence, apart from Adams's help in managing the origination issue, the executive had little impact on this outcome in the following legislative year.

Summary of political process models.   The conclusion is that no single model is useful for predicting outcomes. Indeed, each one plays some role, but all are required for a complete analysis. This is at once disconcerting and reassuring; disconcerting because one expects more predictive power from theoretical models than is possible, and reassuring because the complexity of the policy process requires equally sophisticated analytical tools. The congressional behavior model is best for predicting how the steps are carried out, and the interest group model is best for understanding the intensity of concern among its players. The bureaucratic politics and the interbranch models appear to be the least helpful.

Reid puts the issues into a much less theoretical and more practical framework when he lists four factors that determine the fate of bills in Congress: "policy, personality, parliamentary procedure, and politics." The passage of this water bill certainly had all of them: good public policy, the dogged determination of Senator Domenici to prevail, the parliamentary wizardry of Senator Long, and the ever-present politics. It suggests that good luck is also an important factor in arriving at satisfactory policy outcomes. The saga of this bill suggests that changing federal water policy by legislative means is not a task for the fainthearted.

The Performance of Interest Groups in Recent Water Legislation

Three recent laws are key to understanding the role of interest groups in developing current federal water policy: PL 99-662, the Water Resources Development Act of 1986, PL 99-399, the Safe Drinking Water Act amendments of 1986, and PL 100-4, the Clean Water Act amendments of 1987. Between them these laws cover water resources, ambient water quality, and drinking water quality. To a lesser extent

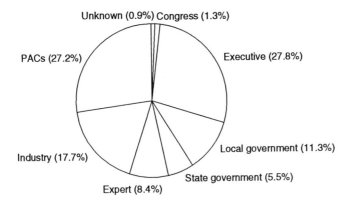

Figure 7.1
Percentages of testimony given to the House Subcommittee on Toxic Substances and Environmental Oversight, 97th Congress, concerning proposed Safe Drinking Water Act amendments.

they also deal with groundwater, wetlands, non-point source pollution, and marine oil spills. Each new law represents many years of intergroup, interagency, and interbranch negotiations.

It is hard to quantify the input of various interest groups by direct observation, but the voluminous legislative history of each bill as it moved through the various committees and subcommittees of Congress is a rich source of information. An admittedly highly simplified approximation of the contribution of each group is the number of pages of hearing testimony it offered. Clearly the actual content of what was said, how it was delivered, and by whom are important, along with whether the committee members were present or absent, asleep or awake, and whether, as one expert pointed out, the testimony took place "before or after the network television crews had left for the day." Given all these aspects of valuing the testimony, the simplistic page counting method has merit precisely because it avoids all value judgments. Moreover, for the Safe Drinking Water Act amendments, the Clean Water Act amendments, and the Water Resources Development Act, page counts appeared to predict the outcome of the legislation.

For the Safe Drinking Water Act amendments the percentages of the testimony given during the 97th and 98th Congresses to the House Subcommittee on Toxic Substances and Environmental Over-

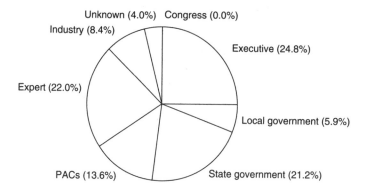

Figure 7.2
Percentages of testimony given to the House Subcommittee on Toxic Substances
and Environmental Oversight, 98th Congress, concerning proposed Safe Drinking Water Act amendments.

sight are shown in figures 7.1 and 7.2, and those for the testimony
given during the 98th Congress to the Subcommittee on Health and
the Environment of the House Energy and Commerce Committee
are shown in figure 7.3. In each case the executive agencies (largely
the EPA) provided the largest amount of testimony, closely followed
by the environmental interest groups (lead by the Natural Resources
Defense Council), with the industry representatives trailing far behind. Since the outcome of this legislative work is widely viewed
as a very strong pro-environment law, this simple analysis provides
a prima facie suggestion that an iron triangle composed of the
EPA, the environmental lobby, and the concerned legislators, was at
work.

The Clean Water Act amendments were viewed by the White
House as excessively environmental and "budget busting" and hence
were vetoed by the president in 1986. The veto was overwhelmingly
overridden (86 to 14 in the Senate and 401 to 26 in the House) as the
first action of the 100th Congress in 1987. Figure 7.4 shows the percentage of testimony given by the interest groups during the 98th
Congress to the Subcommittee on Environmental Pollution of the
Senate Environment and Public Works Committee. Here the bulk of
the testimony was from the environmental interest groups, followed
by the executive agencies; industry, surprisingly, provided only 7.4
percent of the testimony. The environmental iron triangle seemed to

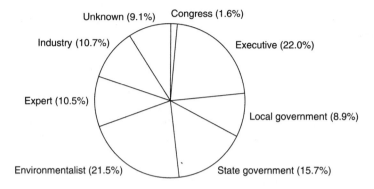

Figure 7.3
Percentages of testimony given to the House Subcommittee on Health and the Environment, 98th Congress, concerning proposed Safe Drinking Water Act amendments.

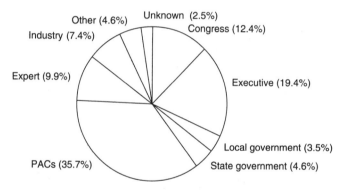

Figure 7.4
Percentages of testimony given to the Senate Subcommittee on Environmental Pollution, 98th Congress, concerning proposed Clean Water Act amendments.

be alive and well in the area of ambient water quality policy even though the executive agencies in principle were taking the anti-bill White House view; their professional biases, and their need to continue working relationships with the environmental groups, would presumably carry them in the opposite direction.

With the economic stakes very high it is surprising that for neither of these bills did industry participate heavily in the hearings process. It could be that industry had already conceded these bills to the opposition, and did not seriously try to influence the outcomes through the hearings process. It could also reflect skill and coordination among the environmental groups that allowed them to mobilize their supporters more easily than industrial groups. It could also reflect the power of congressional staffers in organizing the hearings and issuing invitations to testify.

For the Water Resources Development Act of 1986 the results are much more intriguing. This area of water policy is the area of raw congressional power; this was the first omnibus water bill initiated by the administration rather than Congress since the civil works program began in the last century. One critical review of the outcome of the legislative process (Cortner and Auburg, 1988) saw this bill as the output of one of the most powerful and long-lived iron triangles. In this case the executive agency is the U.S. Army Corps of Engineers, the interests are groups such as the Rivers and Harbors Congress (once the most powerful lobby in Washington), and the local politicians are the eager third vertex. The 1986 bill for the first time seriously incorporated the concerns of environmental groups through such novelties as regulation of groundwater use, but especially through cost sharing and user charges. This was a conscious and successful effort on the part of the Reagan administration and environmental interest groups to cooperate actively. The executive branch was interested largely in scaling down the size of projects and hence their drain on the federal budget. The environmentalists were interested in simply cutting back the projects.

The objective record, interestingly enough, does not record this collaboration. The environmental groups, except for the Wildlife Management Institute, were not represented in the major committee hearings. For example, figure 7.5 shows the distribution of testimony to the Subcommittee on Water Resources of the Senate Public Works Committee on January 24 and 25, 1984. Industry accounted for 78 percent of the testimony and the remainder was from pro-legislation executive branch agencies and legislative branch politicians. Figure 7.6 shows the configuration at the September 10, 1985, hearings of the Senate Finance Committee when the fuel tax and cargo tax provisions were debated. Again industry led the charge, this time against the tax provisions, with the remaining testimony from state and local government and pro-industry legislators. No environmental group is recorded as giving testimony. Similarly in the September 5, 1985, hearings of the House Ways and Means Committee, the bulk of the

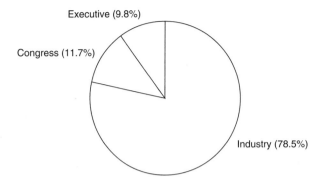

Figure 7.5
Percentages of testimony given to the Senate Subcommittee on Water Resources, 98th Congress, concerning the Water Resources Development Act.

testimony was against the tax provisions of the bill. One environmental group, the Wildlife Management Institute, submitted testimony at this hearing. The silence of the environmental community, on a bill that they were expected to oppose vehemently, was puzzling. The puzzle is resolved when we learn what the record does not show. This is the determined effort of one executive appointee, Robert Dawson, who was intimately associated with the proposed legislation as Assistant Secretary of the Army for Civil Works before moving to the OMB as Deputy Director for Natural Resources.

From a political point of view, Dawson wanted to see the Water Resources Development Act enacted because there had been a hiatus of over 10 years since an omnibus water bill had passed Congress and the construction agencies were hurting badly. He also wanted to limit the federal share of the financial costs of the act by getting a strong cost-sharing provision included in the bill. Dawson saw that the way to achieve this was to immediately engage the environmental community in a dialogue and convince them that some bill was inevitable and that they could get a much preferable version by supporting the cost-sharing aspects of the proposed bill and not worrying about the $16.5 billion in new projects, since they would not go forward if the strict cost-sharing provisions went into force. Dawson was able to persuade the environmentalists not to exercise their de facto veto. In return for their restraint, the environmentalists acquired the (probably permanent) benefit that henceforward all water projects would

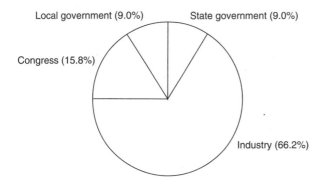

Local government (9.0%)   State government (9.0%)

Congress (15.8%)

Industry (66.2%)

Figure 7.6
Percentages of testimony given to the Senate Finance Committee, 99th Congress, concerning the Water Resources Development Act.

have substantial costs to their beneficiaries, which is certain to reduce the demand for projects by the groups that have historically and very successfully clamored for them.

Political Options for Improving Water Policy

The political process, perhaps more than the other imperatives, seems to be able to break its own rules. It is labyrinthine, labor intensive, and at moments seems to wander far from the subject at hand, appearing highly illogical. But these are the other sides of its extreme flexibility and capacity to reach considerations that may be emotional and almost subconscious, and to give such feelings and traditions a weight in relation to straight economic concerns. The political process constantly weighs such imponderables as the value of the present against the future. In the world of water policy, the political process is the ultimate stage in a movement built up through technical, economic, or even physical developments. Politics socializes those efforts, engaging them in all the mysteries of society and all the actions of human beings.

   The use of the political process to bring about changes in water policy is clearly fraught with many difficulties. As the Domenici bill demonstrates, changing policy by introducing legislation can be done, but not easily. It may be a strength of the system that once legislation is passed it is very difficult to reverse the process, but in settings where the original motivation for the legislation has changed this is a major weakness. For example, the relative emphases on emission controls and ambient quality of the Clean Water Act of 1972 were probably appropriate when originally passed, but are no longer relevant since implementation of the act itself has made other approaches, such as non-point source control, much more appropriate. It is now extremely difficult to make corrections since many iron triangles have sprung up around the existing legislation. The danger is that we will always be fighting yesterday's battles.

Institutional Needs and Possible Responses

One cannot overestimate the tremendous influence of institutions in delineating the boundaries of federal water policy.[1] Long-established organizations, with time-tested outlooks and ways of operating, provide the matrix within which water policy is articulated. Unfortunately, they also serve as constraints upon the development of creative policy options. Successful policies are often impeded not by lack of resources or technology or an insufficiency of economic approaches, but rather by lack of coordination.

Institutional Needs for Coordination

The large number of institutions, committees, and interest groups with legitimate interests in water policy can easily lead to paralysis of the policy process. Beyond those that exist now, many kinds of institutions have been attempted and discarded over the years.

Why has it been so difficult to establish an enduring agency to provide overall coordination to federal water policy? Part of the answer rests with the constitutional separation of powers, and part with the inevitable competition (of authority and prerogative) between the executive and legislative branches. When sizable investments are being made, legislators do not wish to give the executive branch the power to decide which projects get authorized and funded. Within the executive branch, functional agencies prefer to determine their own priorities without centralized direction, and the authorizing committees of Congress are equally interested in evaluating water projects independently. Another part of the answer is the absence of incentives for improved coordination. In the past there has been little reward in being a champion of economy, efficiency, and evenhandedness in federal water investments and regulation. Consequently, little if any political capital has been expended toward such ends.

The interagency nature of the coordinating instruments attempted in the past has also been part of the problem, since members tend to perceive themselves as representing their own agencies instead of the larger national interest.

Executive agencies: the Water Resources Council. An important attempt at consolidation occurred in 1965 during the Johnson administration. Title I of the Water Resources Planning Act created a Water Resources Council composed of the Secretaries of Agriculture, Army, Commerce, Housing and Urban Development, Interior, and Transportation, and (later) Energy and the administrator of the Environmental Protection Agency, and chaired by the Secretary of the Interior. The new federal water agency was to be exclusively intragovernmental and interagency in character. The council's statutory responsibilities ranged from broad policy recommendation and review to specific planning and development activities. The legislation also gave the council a central role in coordinating seven federal-state river basin commissions, and a new program of grants to encourage state water resources planning.

By the mid-1970s it appeared that the council was still not the right device for a comprehensive approach to water resources at the federal level. The General Accounting Office (1977) reported that the council had made only limited progress in carrying out its responsibilities under the Water Resources Planning Act. For example, the attempt at integrated planning was proceeding slowly and was generating products of questionable usefulness. Other critics noted that having a cabinet officer chair the council subjected the agency to undue member bias. In the absence of consensus on a role for the council in policy formulation, it could only follow the path of least resistance among its members. Even the long-awaited innovation of federal-state river basin commissions had come in for its share of criticism. While some commissions had proved useful, many were simply not appropriate to state needs. And the principle of state representation coequal with federal representation that the basin commissions embodied was not carried forward at the level of the council's own deliberations.

The ultimate difficulty in finding an appropriate organizational format seems to be the simple matter of perception. There are as many versions of federal water policy as there are experts, participants, and claimants. Similarly, the Water Resources Council, even with its explicit statutory mandate, lacked real focus. At least four different roles were visualized by its proponents: as an independent evaluator and formulator of federal water programs for the president and Congress; as a centralized operation for data gathering, resource assessment, and research; as an administrator of technical assistance programs for federal agencies, states, and local jurisdictions; as an agency charged with setting and enforcing standards for federal and/or state plans and programs. Without a consensus on its goals, and with no clear line of accountability within the federal establishment, the council drifted for the decade and a half of its existence doing the best it could. Attempts at making things better ended up making them worse. President Carter's 1979 executive order giving the council independent review authority over agency projects was almost immediately rescinded by Congress. Finally under President Reagan, two years later, the executive and legislative branches joined forces to zero-fund the council and the river basin commissions, effectively ending their existence.

But the initiatives of the Reagan administration fared no better. The cabinet-level Council on Natural Resources, created in 1982, proved ineffectual at coordination of water policy and became essentially nonfunctional after its principal enthusiast, Interior Secretary James Watt, left office. Equally short-lived was the Office of Water Policy, the purported successor to the Water Resources Council. The only remaining vestiges of the Water Resources Council today are the Principles and Standards (renamed Principles and Guidelines, U.S. Water Resources Council, 1983) administered by the Office of Management and Budget. In fact, by default the OMB now serves as the principal agency in the executive branch concerned with water resources coordination, a very inadequate situation, both for the hopelessly overloaded OMB, whose responsibilities extend over the entire range of U.S. government activities, and for the needs of the water resources field.

Department of Natural Resources.   At least six presidents have tried to formalize interagency cooperation in water and natural resources. It was not until the late seventies, however, during the Carter administration, that a much more radical proposal than consolidation surfaced and was almost implemented—a Department of Natural Resources charged "with the mission of managing the Nation's natural resources for multiple purposes, including protection, preservation and wise use" (Office of Management and Budget, 1978, p. 1). In August 1977 the president directed the Reorganization Project, an office created within the OMB to fulfill his campaign promise to improve federal efficiency, to review all federal programs for managing natural resources and protecting the environment. The Reorganization Project reported:

The federal institutions . . . were designed one by one in a previous era to address specific problems. . . . New organizations were established . . . for narrow purposes such as the encouragement of settlement of areas of the public domain, providing irrigation water to farms in arid areas, managing and conserving timber, or locating and developing energy and mineral resources. Over time, as the Nation's needs changed, these scattered institutions have been given broader responsibilities by Congress. This has evolved into a system that lacks comprehensiveness and, at the same time, contains overlaps. It produces delays, unnecessary confusion and inconsistencies, excessive costs, and narrowly based decisions. (Office of Management and Budget, 1978, p. 2)

The Project presented the president with several options, including the creation of a Department of Natural Resources. This new department would have included all functions of the present Department of the Interior, in addition to taking over the U.S. Forest Service from the Department of Agriculture and the National Oceanic and Atmospheric Administration from the Department of Commerce. It would also have included the Water Resources Council, while giving up the construction functions of the Bureau of Reclamation and of the Soil Conservation Service's watershed program to the Army Corps of Engineers. The Department of Natural Resources would have had a staff of 89,800 permanent employees and an annual budget exceeding $7 billion, but it was estimated that $100 million per year and 2,000 permanent jobs could be saved by this reorganization.

A press conference was scheduled for March 1, 1979, to announce the reorganization. However, after an evening meeting on February 28 at which Secretary of the Interior Cecil Andrus detailed the intensity of congressional and agency opposition, President Carter canceled the news conference; on May 15, administration officials quietly announced that plans for a Department of Natural Resources were being dropped.

Institutions for Strengthening the States' Role

Since states are sovereign over water, it may appear strange to talk about "strengthening the states' role." But the federal government has nibbled away at the states' role so assiduously that there is a widespread desire to reverse this centralizing trend. The government has, over time, gone well beyond its role of establishing a framework of national objectives and standards in consultation with the states. Redressing this was initially a goal of the "New Federalism" of the Reagan administration, but when Interior Secretary Watt and EPA Administrator Anne Gorsuch plunged into dismantling the federal resources responsibility, their effort was seen as a direct assault on federal regulation with no compensating strengthening of the states' role. With their resignations, the main federal water regulatory agencies were able to pursue more balanced roles vis-à-vis the states.

A major development was the devolution of funding for wastewater treatment facilities from the federal government to the states under the Clean Water Act amendments of 1987. This legislation recognized the responsibility of the federal government in achieving water quality goals by helping in the funding of municipal facilities, but only for a limited time. The federal share was reduced from 75 to 55 percent of the capital costs, and subsidies were to be eliminated entirely after five years.[2] The federal government has also devolved many of the powers of the National Pollutant Discharge Elimination System (NPDES) to the states. By the end of 1992 more than half the states were issuing their own NPDES permits for wastewater discharges by municipalities and industry.

While these developments represent significant steps toward the return of sovereignty over water to the states, the federal government has kept the right to set the overall quality standards for effluents and drinking water. Since there is little room for adjusting federal regulations to local conditions, the result looks like a poor deal to many state officials who see federal-state sharing as essentially meaning "we regulate and you pay."

Another initiative of the Reagan administration was to strengthen the cost-sharing requirements in the Water Resources Development Act of 1986. Localities are now required to pay an increasing proportion of the cost of developing water resources projects, in return for which the president allowed Congress to authorize a larger number of water projects. Technically this allowed the states to act as partners with localities within their state in order to cofinance projects.

From the states' side, at the urging of a special water resources committee, the National Governors Association in 1987 adopted a set of principles to be applied to national water policy. Water management should be the primary responsibility of the states and their delegated interstate agencies, but it should be approached in a more comprehensive and coordinated manner at all levels. Improved continuity was called for in federal support for water programs; also greater consistency with state and interstate plans and programs. The entire federal support system should be applied more flexibly, the governors observed, and criteria for federally assisted projects and programs should be refined and applied uniformly. Among special issues singled out for heightened federal attention were water conservation and water resources research.

Institutional Needs for Resource Assessment, Research, Planning, and Management

The search for research coordination has not been particularly fruitful over the years. In 1962 the President's Council for Science and Technology established a formal Committee on Water Resources Research comprised of representatives from the federal water resource agencies with research missions. The committee persisted until 1977 but made little progress either toward a common national water resources research agenda or toward effective interagency coordination. Subsequent reviews of water research management[3] all concluded that the existing system was inadequate, and that the inadequacies were not incidental but rather a function of the dispersion and the political context in which such agencies usually operate. Nonetheless, the system continues.

In 1964, an attempt was made to provide explicitly for water resources research outside the mission agencies. Congress enacted the Water Resources Research Act (PL 88-379), modeled after the Hatch Act of 1887 that authorized state agricultural experiment stations. Under Title I of the 1964 act, a state university in each state could establish a water research center with federal funding of up to $100,000 annually. Title II provided an additional, competitive annual grants program in water research of $1 million per annum open to virtually any entity with appropriate research capabilities.

In subsequent reauthorizations, Congress tinkered with the research machinery it had created. By 1978 it had expanded the water resources research mission to include technology development and transfer and regional research consortia. To make research more relevant, Congress required closer consultation among federal agency officials. Increasingly strong directives were given to the president and the Secretary of the Interior to coordinate federal research spending. But by 1984 disenchantment with federal water research was so pervasive that the program of water resources research institutes was zero-funded in the administration's budget. It survived only through a congressional override of President Reagan's veto of the institutes' 1984 reauthorization act.

Part of the problem was indecision on the part of the Department of the Interior. In early years, a formal Office of Water Resources Research (OWRR) was created within Interior to take responsibility for the program. In 1974 OWRR was merged into an Office of Water Research and Technology (OWRT). When OWRT was abolished in 1981, oversight shifted to a short-lived Office of Water Policy within Interior, which became moribund in 1983, and then to Interior's U.S. Geological Survey. The state-based research institutes similarly suffered from federal ambivalent federal interest and support. During the 101st Congress, the nation's water resources institutes funding was secured through 1994 (PL 101-397) at the level of $100,000 per institute. They are clearly facing an uncertain future; each year OMB asks for a 51 percent reduction and, so far, each year Congress has put the money back in.[4] This is ample evidence that a mission-oriented federal agency with research programs of its own may not be the safest choice to manage a tangential, largely external research effort.

At the same time, other developments have been substantially reinforcing and reshaping the environment for water resources research nationally. Among them are the Safe Drinking Water Act amendments of 1986, which require that all public water supplies meet prescribed contaminant standards; the Clean Water Act amendments of 1987, which shift the financing burden of meeting stream quality standards to state and local jurisdictions; the new cost-sharing requirements of the Water Resources Development Act of 1986; and the new ecosystems concerns of the 1992 Reclamation Projects Authorization and Adjustment Act. At no time have research needs been more compelling, nor spread over more aspects of water use, management, protection, and development. It is hardly the occasion to have a large part of the nation's water research establishment on the brink of bankruptcy.

Taken together, recent developments seem to constitute a promising new context for water resources research—but one that needs to be geared more to state and local actions. Thus, a reauthorization of the Water Resources Research Act takes on added meaning and represents an opportunity to articulate an entirely new federal role in research. What should the basic elements of such a program be?

First, all the research avenues that exist now are still needed, but ought to function more as networks than as isolated programs, in order to deliver research and research products more effectively. Second, there should be a closer relationship between researchers and policymakers so that research will inform policy more directly and with less delay, and so that policymakers and water users help direct research. Third, the need for research coordination remains acute, both horizontally (e.g., among the federal departments) and vertically (between the mission agencies and institutions at state, local, and nongovernmental levels that perform or consume research). Fourth, the principal impediments to optimal water management and use are as often social and political as they are scientific and technological. A significant gap exists between the real nature of water resources problems and the almost exclusively engineering and physical science bent of the present research community.

Funding of research.   How should water research be funded? The needs are certainly substantial (in 1987, for groundwater alone, more than $215 million was spent for research and technology). At a time of substantial federal deficits, the prospect of support from general treasury funds is not promising for the heightened level of research needed. With that reality in mind, other sources should be explored.

In principle, funding for water research should first be drawn from polluters and beneficiaries of federally subsidized water (e.g., municipalities, industries, and irrigation and navigation interests). In fact, the closer the financing responsibilities are to those who benefit from the research, the better the work will be. But other options should be explored too. For example, the modest private research allocation ($4 million annually) derived from voluntary contributions to the American Water Works Association Research Foundation contrasts starkly with the more than $330 million raised annually by the industry-wide Electric Power Research Institute. Like their electric

utility counterparts, the water industry would be well advised to use
research to spread the cost of developing new approaches and technologies, develop and expand cooperative research mechanisms of
their own, and learn to build fruitful research and extension partner-
ships with their academic and governmental counterparts. It could
also use a similar utility-wide taxing scheme; a 1 cent per 1,000 gallons tax could raise as much as $130 million per year for research and
development purposes.

Who collects, stores, and disseminates data? Good water decisions, like good decisions elsewhere, are heavily dependent on the
availability of adequate and timely information. Approximately 1,100
water-related information sources are available throughout the country, and no major information gaps are apparent. The bad news is the
state of the current information distribution system. There is widespread ignorance of which data are available, from whom, and where,
and even when known the sources are often not readily accessible. In
short, the present system falls substantially short of effectively dis-
tributing the information it has.

Over time, the collection of water resources data has attracted
considerable attention. The early emphasis was on simply establishing adequate data bases. As early as 1950, for example, the Cooke
Commission called for systematic geological and hydrological surveys to gather information about surface and groundwater resources
sufficient to enable 10- and 20-year forecasts of water uses and requirements. In 1955, the President's Advisory Committee on Water
Resources Policy recommended the establishment of a basic data program to assist in water resources planning.

By the time of the Senate Select Committee on National Water
Resources report (1961), the emphasis had begun to swing away from
collection to application—what should be done with the information
assembled. The Kerr Committee recommended biennial assessments
of the nation's water resources, expecting significant shortfalls due to
rising demands.

The federal Water Resources Council came into being under Title I of the Water Resources Planning Act of 1965. A statutory function of the council was the preparation of biennial national water
resources assessments. The National Water Commission in 1973 recommended mergers of individual agency efforts to achieve a comprehensive, consolidated data system; a sorting out of responsibilities
between the federal government and the states; and a uniform collec-

tion, storage, and retrieval program. The emphasis was on data management, not just data collection. The commission called for the council to install a formal water data referral center to serve as the information hub for a network of state and federal participants and to oversee the mushrooming federal data collection programs.

Regrettably, few of these sensible suggestions were followed. The proposed water data referral center was never established, and in the course of 15 years only two national assessments were produced, neither one very satisfactory. Without the focus that the center and biennial assessments would provide, the data that was collected and analyzed was not always appropriate. Increasing dissatisfaction with the process led Congress to low appropriations, delays, and a gradual erosion of commitment to an integrated data management program.

By the time of the council's demise in 1981, the stage was set for alternative approaches. Federal line water agencies had begun filling the vacuum with their individual data gathering and evaluation efforts, often with the explicit blessings of Congress. The Office of Management and Budget stepped into the breach with Circular A-67, a directive affecting the water data acquisition activities of some 28 federal agencies. And after the failure of Interior's Office of Water Policy (the successor to the Water Resources Council), the Reagan administration called upon the U.S. Geological Survey to establish an Office of Water Data Co-ordination to carry out Circular A-67's plan. This was encouraging, but there were glaring omissions in the agency programs included; the most notable absentees were the meteorological data gathered by the National Oceanic and Atmospheric Administration (NOAA), the upstream sedimentation and snow survey data collected by the Soil Conservation Service, and the water quality information furnished to EPA by the states. The integration of information sources envisioned by the earlier policy commissions had taken a back seat to a variety of individual data efforts.

At present, three of the principal federal data bases are administered by the Water Resources Division of the Geological Survey (Interior). They assess water quality through fixed station monitoring, reconnaissance during periods of special conditions (storms or floods), and intensive cause-and-effect studies on specific problem reaches of rivers and streams.

Two additional USGS data bases are designed to facilitate storage and retrieval generally. One is the National Water Storage and Retrieval System, a repository established in 1971 for streamflow and groundwater data and the results of chemical, physical, biological, and radiochemical analyses. The other is the National Water Data Exchange, designed to facilitate access to a variety of water-related data bases in and out of Interior. Despite the sizable investment in such data systems, the USGS effort appears to have significant weaknesses. For example, it does not monitor trace organics, biological indicators, or pesticides, and overlays EPA operations under Section 305(b) of the Clean Water Act.

The EPA has developed a water data storage and retrieval system of its own. Called STORET, it contains water quality information for surface and groundwater received from several federal agencies, more than 40 state agencies, and the current USGS-state cooperative investigations program. Currently the EPA claims that water quality and quantity data are available for more than 750,000 sites around the country.

The third piece of the water data puzzle is administered by NOAA, which manages the meteorological information so central to the entire hydrological cycle. NOAA's Environmental Data Service collects climatological data, and its parent the Department of Commerce bears federal responsibility for providing floodplain information and flood warning services. Typically, NOAA has its own data system, the National Environmental Data Referral System. Regrettably, no direct linkages exist between the EPA's system and NOAA's system, nor between NOAA and the USGS. Steps are now under way to provide some automatic switching capabilities between the data bases administered by the USGS and the EPA.

Institutional Needs for Financing and Operating Water Resource Systems

Improper pricing of water, rather than inadequacy of supply, is most often the chief culprit in water shortages. Charging users what water really costs to produce, manage, treat, deliver, and dispose of after use is so sensible a concept that it hardly merits elaboration. But the fact is that few water delivery systems follow such a policy today. If we have a national water problem, it is largely due to this, and is largely of our own making. What should be done to remedy misallocations, to take advantage of recognized market efficiencies, and to improve our national recognition that water, rather than a free good, is a commodity with substantial economic as well as social value?

Several preconditions are necessary before viable water markets can exist (see chapter 6). Water has to be available, of course. Equally important, there has to be the technology and/or infrastructure necessary to move that water efficiently to users. Suppliers must hold property rights in the water sufficient to transform it into a commodity that can be bought and sold. Participants in the exchange must see sufficient value in the transaction to give up water for something else, and give up something else for water, and transaction costs must be low enough not to erase this margin of benefit. Exchange arrangements need to be sanctioned (or at least not opposed) by existing social, political, and administrative institutions. These conditions are being met in many water markets in the West.

But the market approach is far from well accepted in the United States of today. It can run afoul of the pervasive concept that water should be freely available to all, and that in-stream water for fish and wildlife, in particular, should be provided at no cost. The practice of making a profit from water sales, or even building a sinking fund from revenue surpluses in anticipation of future needs, is foreign to many purveyors, offensive to many of their constituencies, and frequently illegal. And proper water pricing can encounter a firestorm of political protest, since the ultimate step is apt to be at least a doubling of prevailing rates.

Students of water market theory have concluded that what is needed most is neither a flat prohibition against water transfers nor carte blanche to sell water, but some sort of orderly process with standards to ensure that transfers are equitable. Each transaction should provide the least-cost source of a new supply for the user. After consideration of social costs, it should provide a net increase in social benefits. It should account for the concerns of parties not directly involved in the transfer, such as the downstream interest. And it should leave no one worse off. To guarantee the latter, conditions and limits may have to be placed upon the transfers, such as volume ceilings, or time and seasonal restrictions. Provisions may have to be included for mitigation or relief from injury; and a process may have to be set up for negotiation or adjudication of claims. Because of the importance of third-party effects, such as environmental impacts, it seems unlikely that water markets can function equitably in an entirely unregulated form.

Federal water project entitlements seem to provide one of the largest and most immediate opportunities for the reallocation of water.

The Bureau of Reclamation, for example, now supplies roughly a quarter of the water used in the West. The resources under Reclamation control are long overdue for reallocation, but similar opportunities may be at hand in other water project agencies—the Corps of Engineers, the Tennessee Valley Authority, and the Department of Agriculture's Soil Conservation Service, for example. There exists an apparent potential for project redevelopment throughout the federal system as a means of achieving market-sensitive water reallocations, acquiring new sources at modest or even marginal costs, and raising badly needed water revenues.

Viewed historically, federal water project development has proceeded unevenly, inefficiently, and inequitably. It has been driven largely by the dictates of distributive politics. The result has been water often not available where it is most needed or desired and wasted or abused where it is available. Water in federally developed projects is generally allocated for fixed purposes (e.g., flood control, hydroelectric power, irrigation), with a hierarchy of uses established even in multipurpose projects. Despite an exhaustive legislative legacy of local contributions of land, easements, rights-of-way, and maintenance, real financial cost sharing has only recently become an accepted practice, as in the Water Resources Development Act of 1986. At that, cost sharing has been applied only to certain federal agencies and at varying rates. Because water cannot be sold simply to those most willing to pay for it, or in many cases because this principle has been overinterpreted, many federal projects have come to require significant subsidies, not just for their initial construction but for ongoing operations as well.

The historical reluctance of the federal government to take a market approach to water has had a ripple effect at virtually every other government level. It is one reason for the unwillingness of water supply agencies, for example, to charge for drinking water on a full marginal cost basis. It also in large part created the pressure from state and local governments for massive infusions of federal aid to help with water-related programs. In certain aspects the practice of nonmarket approaches has reached the level of absurdity. In parts of the West, for example, highly subsidized water is being used to irrigate price-supported crops currently in surplus whose waste products so pollute groundwater and endanger wildlife as to require federally funded pollution control measures (e.g., California's Kesterson Reservoir). The place to begin is with a reappraisal of the prior federal investments in water projects and a reallocation of the output of such projects to achieve greater efficiency and more appropriate uses in the future.

The response to drought conditions in the Potomac River region in the mid-1960s and the late 1970s is one specific example of the potential of new initiatives. The major political forces in the region (federal and state agencies, local governments, and water supply authorities) agreed to explore joint approaches to increasing water supplies during periods of low flow in the river. Computer simulation revealed that a coordinated system of releases, withdrawals, and uses would obviate the immediate need for major new structural facilities in the basin. Rational reallocation of the river's already available resources turned out to be the most timely and cost-effective solution to the problem (Sheer, 1986).

For existing federal projects, there is a need for a systematic reexamination of each physical facility where water is already stored, or where the potential for storage exists. Analysts should question the relevance of present uses, consider the alternatives available to meet original purposes that are still valid, calculate the extent to which prior investments and carrying costs have been recovered, look for opportunities to modify the facilities or activities to meet new or foreseeable needs, and review the extent to which the revenues and/or fees collected are reflective of current market conditions. The analyses should be done uniformly across the federal establishment and without regard to contractual commitments, agency policy, or statutory law. The touchstone should be whether or not each project has achieved its full potential.[5]

There are significant discrepancies in many federal projects between the subsidized rates that local beneficiaries pay and the market rates for the same commodities such as irrigation water, hydropower, and recreation. Even a partial rectification of these skewed policies could provide significant increments of water and funds. The increasing federal role in groundwater research and management may warrant special revenue-raising devices, such as a federal tax comparable to the growing number of state wellhead pump taxes. Federally mandated surface water discharge permits (now 65,000 in number for EPA alone) represent a potential revenue source. Taxes on contaminants seem more than justified—particularly on pesticides and fertilizers that contribute so much to the non-point source problem nationally. Contamination taxes could be recovered readily through the price structure of the crops grown, by a levy on the agricultural chemicals themselves, or by taxes on the land where the chemicals are used.

Where federal water creates a benefit, the federal government has a claim. Much more could be done with charges imposed directly for the use of federal resources or facilities, or surcharges placed upon the equipment or facilities needed to enjoy the resources (e.g., boating or fishing equipment). But perhaps the greatest potential for revenue generation at federal facilities lies in the areas of municipal and industrial water supplies, especially in the high growth areas of the West and Southwest and the metropolitan portions of the East. A ready market for water exists in these regions at prices that would be fully competitive. The opportunities for revenue enhancement throughout the water domain appear very substantial.

If federal water revenues do indeed become substantial, some sort of institutional depository would be in order. A national Water Trust Fund, analogous to the Highway Trust Fund, could receive revenues from a variety of sources and hold them subject to appropriation. A series of subaccounts could be created to retain the integrity of the present Inland Waterways and Harbor Maintenance Trust Fund and other earmarked sources. Over time, such a Water Trust Fund should become the primary source of funds for water resources operations, research, development, and protection carried out by or through the federal government, replacing the more than $5 billion now appropriated annually from general revenues.

Financing.   The generic term *infrastructure* usually refers to the physical components of services provided by public works: roads, ports, bridges, rail transit systems, and the overhead lines of electric utilities. The unseen infrastructure, however, is even more pervasive. Human communities and commercial operations are served by a vast network of underground pipelines, mains, sewers, wires, and conduits. The term infrastructure can also be used to describe the institutional and other organizational attributes of these public works systems. Public spending for infrastructure facilities is estimated at more than $100 billion annually, a sum that represents nearly 7 percent of all governmental outlays.

However, it would be a mistake to think of the infrastructure problem as one of dollars alone. It is also a problem of institutions. Institutions must be in place to guarantee the preservation of the capital assets created. If an infrastructure bank is established, for example, it would be entirely appropriate to condition the financing of projects upon the imposition of a sensible pricing structure, or a measure of consistency in the provision of facilities on an areawide basis. Federal requirements of this sort could help communities solve the difficult political problem of regional equity in a full user-pays approach.

Over the next decade, more than $100 billion will be needed to construct and maintain the water supply and the water quality infrastructure. There is an absolute need for government financing for only a very limited number of goods or services—the so-called public goods that everyone enjoys and nobody owns. In the water area these are an extremely limited part of the total financing requirement. The navigational aids for which the Congress first appropriated money in 1789 remain the classical examples, but even these can now be financed by the Inland Waterways and Harbor Maintenance Trust Fund which is replenished by a tax on fuel used on federal waterways. In fact, close examination discloses no goods or services produced in the water or water quality area that do not have some beneficiary who can be taxed or otherwise charged for the benefits received; even instream flows can be charged to fishery and recreation interests. This was not always true, but with the extension of the federal water quality laws to the "waters of the United States" the externalities of production that generate water pollution are now fully regulatable and taxable.

With removal of the absolute need for federal funding for water policy implementation, the problem now becomes finding the most efficient way to achieve the clearly enunciated federal water goals. Some analysts argue on social equity grounds that the federal government is still the best source, but there is serious doubt about the efficiency of federal financing. In fact, studies show (Congressional Budget Office, 1983) that communities tend to spend up to twice as much capital on federally aided sewage treatment plants as on locally financed ones. Requests by local communities to scale down projects when they discover that the Corps of Engineers is seriously insisting on cost sharing make the same point. The revisions suggest that the original plans were often inflated beyond the rationale of pure economic efficiency.

The history of federal involvement in water development reveals some approaches that continue to have some merit. The most important concept is that of establishing a special fund from which the capital and maybe the operating costs can be met. With the founding of the Reclamation Service in 1902, a Reclamation Fund was also established to pay for irrigation projects. The fund was initially capitalized by the sale of public lands and was intended to be self-sustaining, with the farmers repaying the service for the works provided. While the concept was an excellent one, the land sale revenue was not sufficient to satisfy Congress for the scale of operations it wished from the Reclamation Service. By 1914 the fund was being replenished by congressional appropriations, which in itself would have been defensible if strict accounting procedures had been followed for repayment by the farmers. The fund, however, soon became a source of federal subsidy for "worthwhile" goals of doubtful relevance. If the original purposes of flexibility, accountability, and program targeting had been followed, the sorry history of federal involvement in irrigation (Reisner, 1986) could have been quite different.

In the case of the Reclamation Fund it could be argued that the sale of land was a natural way to capitalize the fund. What is the equivalent "natural" source of funding for environmental funds? There is little difference between selling property rights to land and selling property rights to the environment. In fact, the formulation of federal regulations on water quality discharges immediately created property rights to pollute by the users of the streams. These rights are currently available essentially without cost (other than a small filing fee) by filing for a discharge permit. (Over 65,000 have already been issued.) A suitable charge could be made for this permit and the revenue used to capitalize a fund.

Other examples are the Waterways Trust Fund established in 1979, funded from a fuel tax on waterways users, and the State Revolving Loan Fund established under the 1987 Clean Water Act amendments with a $9 billion grant from Congress. Other funds, however, like the Airports Trust Fund and the Highway Trust Fund, have increasingly come under the budgetary control of Congress, where they now help defray the budget deficit by not being expended for their stated purposes. The National Commission on Infrastructure also recommended the creation of a federal infrastructure trust fund.

A second important concept is that water is a valuable commodity and funds needed to pay for its administration and development should be raised directly from sale of the resources or of services associated with the resources. This raises the issues of creative financing and privatization. History is replete with all sorts of subsidy and hidden subsidy schemes for financing water and wastewater systems. Many of these were eliminated by the Tax Reform Act of 1985, which closed many of the loopholes in the financing of municipal bonds. For example, it was popular to use local tax-exempt revenue bonds to finance wastewater treatment plants for joint municipal and industrial wastes. A combination of the 1972 Clean Water Act and the Tax Reform Act has essentially closed off that option. Other options such as leaseback of municipal facilities are no longer attractive under the new tax laws.

Other sources of financing could come from directly taxing water use. For example, the Clean Water Council (1990) claims that a 5¢/1,000 gallons tax on drinking water supplies would raise as much as $900 million per year for a trust fund, and a 5¢/1,000 gallons tax on wastewater could add another $750 million per year. There is also the prospect of more widespread participation by the private sector in water supply and wastewater management. Although 29 percent of the drinking water in the United States is supplied by investor-owned utilities, there does not seem to be a movement toward expanding that fraction of the market. In the United States almost all of the major systems are government-owned and managed. Indeed, in some recent cases cities have taken over previously privately owned water companies because of lack of adequate performance. Entrepreneurs in the United States do not believe that water utilities are potentially lucrative economic properties at present because of the regulatory overhang of the unissued standards for many of the contaminants listed in the Safe Drinking Water Act. Until all of the standards have been issued and their economic implications examined, there is little likelihood of large expansion of the private sector in this area.

There is, however, a fresh interest in privatization of wastewater management facilities. This seems to be taking two different forms; for small cities the authorities are leasing out the publicly owned facilities to be managed by private companies, and for larger cities only certain functions are being leased to private companies. The leasing in both cases tends to be limited to the treatment plants themselves and not to the sewer system and outfalls. There have also been significant changes in the public acceptance of marketing schemes for effluents. The Clean Air Act amendments of 1990 introduced limited trading of effluent permits for SO_2 between electric power utilities, and similar proposals are being considered for water pollutants in the revision of the Clean Water Act to be taken up during the 103rd Congress.

Although water facilities appear to be in somewhat better shape than other parts of the national infrastructure, two major problems loom large on the horizon: insufficient funds for renewal and development, and the absence of financing mechanisms for the investments required. Several principles seem useful in clarifying the future federal role in water infrastructure replacement and development.

1. Federal statutes should be far clearer and more rigorous in specifying who is responsible for doing what and when.
2. User cost pricing should be employed to match investment levels with economic demand and help recover costs from those who benefit from the use of the facilities.
3. The federal government should exercise restraint in limiting the revenue-raising capacities of state and local governments for such purposes.
4. Federal financing aid mechanisms should be coupled with the actual infrastructure operations so that leverage can be applied to achieve better local management of physical facilities.
5. Federal assistance commitments should give state and local governments stability over several years to aid in long-range planning, and allow them flexibility in the use of funds.
6. Finally, the rules governing federal assistance should recognize the distinct financial and managerial circumstances facing small, poor, and rural governments, and the nation's traditional concern for social equity. The special problems of El Paso County, Texas, where 350 colonias (suburbs) are without water and sewerage, are a case in point.

Institutional Needs for Protecting the Environment

The Environmental Protection Agency is a misnomer; the agency spends the bulk of its time and effort in protecting human health rather than protecting the environment (Pedersen, 1988). Even though the original 1972 Clean Water Act, PL 92-500, represented a radical breakthrough in shifting the concern of the water legislation from human use to the maintenance of the biological integrity of the nation's waters (the fishable/swimmable goal), the bulk of the regulatory work is still directed toward human health effects. This is nowhere better demonstrated than in the 1986 Safe Drinking Water amendments, which direct the EPA to write standards for, and regulate, 83 contaminants in the first three years. Within the next six years it had to do the same for another 50 contaminants. (It actually achieved 48 by the end of 1991.)

As the EPA's own assessment of its activities indicates (U.S. Environmental Protection Agency, 1987), while its concerns are directed almost exclusively to human health, its activities are not necessarily those with the greatest potential to improve human health. Which aspects of the environment should the EPA concentrate on, if any? In the area of air quality it has been concerned with visibility in pristine areas through nondegradation programs. What are the equivalent programs in the water area? Anti-backsliding (preventing the weakening of effluent limits when a water quality permit is reissued, renewed, or modified) is one provision the agency has consistently used, much to the annoyance of some economists. EPA's administrator under the Bush administration, William Reilly, promoted concerns with large-scale resource and environmental protection, such as loss of tropical rain forests, acid rain, ozone depletion, and global warming. Does the EPA have a comparative advantage in these areas? Is the EPA the appropriate agency to deal with these issues, which are largely international? Probably not. And who is protecting the U.S. environment? At present no one institution has prime responsibility.

A permanent and definitive realignment could look like this:

1. All the health-related research, standard setting, and regulation of EPA are moved to the Department of Health and Human Services.
2. All of the remaining federal water programs are reassigned to the Interior Department.
3. The Army Corps of Engineers is gradually eased out of the resource development aspects of water and becomes a contractor to maintain existing facilities—which is in fact the largest part of the Corps's present civil work.
4. The Bureau of Reclamation retreats to Denver to operate and maintain its facilities and to look for a mission in the area of environmental management in the western states, as it announced in 1987 it would like to do (U.S. Department of the Interior, 1987).
5. TVA is converted into an autonomous service corporation for the TVA infrastructure.
6. The EPA becomes an "environmental protection" agency with a focus on maintaining the ambient environment and ecosystems.
7. A new federal water resources planning and management agency is created in the Interior Department from the relevant parts of Corps of Engineers and the Bureau of Reclamation. This is to ensure that there will still be a significant federal presence in this key area.
8. The Soil Conservation Service is restricted to soil conservation and agricultural runoff problems.

After a major realignment along these lines, the water coordination problems of the federal government would be greatly reduced. Each of the important programs would be under a cabinet secretary. Health-related issues would be under an agency that is charged to look at all aspects of health, not just the ambient environmental aspects. A revised and revitalized Water Resources Council (see next chapter) could easily coordinate all the federal water programs.

Continued Regulation of Water and Water Quality

The Clean Water Act, the Safe Drinking Water Act, the Water Resources Development Act, the various Wild and Scenic Rivers, Wildlife, and Fisheries acts—all enacted in the last decade—mean that the United States has a set of water laws and regulations that are among the most comprehensive and toughest in the world. Apart from groundwater issues, there appears to be no need for additional legislation in new areas. The Clean Water Act and the Safe Drinking Water Act are likely to be amended during the 103rd Congress. Groundwater legislation is critical because this is the last part of the hydrological cycle to be regulated, and the hydrological imperatives require it to be integrated into the pattern of management immediately.

Almost half of the nation's drinking water comes from groundwater, a source that is easily contaminated by indiscriminate and largely unregulated discharges of pollutants and, once contaminated, is very difficult to clean up. During the 100th Congress 18 groundwater bills were filed, and none was passed into law. The 101st and 102nd Congresses saw little legislative action on groundwater, and Congress did not pass any of the bills it considered. Some form of research and coordination with state programs would seem to be the most appropriate legislation for the short run, until the extent of the actual problems is better known. More detailed and interventionary legislation may or may not be indicated.[6]

Non-point source pollution strikes at the heart of many of the policy issues set forth earlier. It involves the responsibilities of a number of federal agencies, and includes uncertainty over the respective roles of federal, state, and local actors. There is substantial overlap in existing legislation that gets into land use issues as well as water quality. Better regulations in this area would also be beneficial.[7]

The Need to Educate Everyone about Water

Persuading Americans that water does not naturally and perpetually come out of faucets and go down drains cost-free is an important task for water managers and specialists in the years ahead. An improved national educational effort in water resources, though clearly needed, is immensely difficult given the pervasive nature of the resource and its many configurations and manifestations across the country. Small wonder that the educational programs are most frequently localized, targeted to particular problems and settings, and sponsored by a wide range of public agencies and private organizations.

What to do about water education is a matter of some debate. To be effective, the message must be tailored to meet particular problems in particular places. But there is growing evidence that awareness is a matter of national concern, and that economies and efficiencies might be obtained through a countrywide educational effort. Like so many other aspects of water, fragmentation of effort is a principal culprit.

Two particularly effective earlier national efforts to raise the public's understanding of water issues may be instructive. The first was the League of Women Voters national agenda item on water undertaken in the late 1950s. The second was the choice of water as the national debate topic for high schools in 1985–86.

In 1956 the League of Women Voters, a civic organization with more than 110,000 members across the United States, selected water resources as a major subject for study and action. The timing was especially appropriate, for the country was in the grip of water issues of various kinds. Water for agriculture and development was of prime concern throughout much of America's heartland and the West. The Northeast was just beginning to recover from the record floods of hurricane Diane. The Watershed Protection and Flood Prevention Act had been passed by Congress, assigning the Agriculture Department's Soil Conservation Service a major new role in upstream areas. And the nation had been promised a comprehensive look at water resources by the Senate Select Committee on National Water Resources headed by Senator Robert Kerr of Oklahoma.

In the mid-1980s, a second exemplary national water awareness effort came into being. At the urging of the active water committee of the Texas Society of Professional Engineers, the National Federation of State High School Associations accepted the management of the nation's water resources as the topic for the 1984–85 national high school debate program. Each year 20,000 high schools nationally take part in the program, involving an estimated 100,000 students (grades 9–12). All of them debated water policy from September to May of the 1985–86 academic year, and continued at summer institutes held at 36 colleges. Counting families and audiences, at least a million people were exposed to the concept of a national water policy and a discussion of current and future water issues.

The entire West has been a leader in the water awareness effort. In the Southwest, emphasis has been on groundwater issues. The Arizona Department of Water Resources produced special materials for the state's public schools, and a 27-minute educational video describing the groundwater overdraft problem. The Texas Society of Professional Engineers put together a school program on conservation; textbook publishers were persuaded to incorporate more information on the importance of water resources.

Elsewhere in the country, there are similar manifestations of a growing interest in accelerating water awareness. The need for water awareness is equally compelling in the large metropolitan regions of the East. Metropolitan New York, long renowned for its high-quality water and advanced system of reservoirs and water supply planning, must now also make large investments to keep the system functioning. It remains one of the few large cities in the country without comprehensive water metering, although in 1989 it became committed to metering multifamily units and the meters are now being installed.

A Blueprint for Water Policy

This book has tried to give a balanced view of water use in the United States and of the federal policies that have evolved for its regulation. Throughout, it has avoided the temptation to take sides in the debates between environmentalists who see the resource in dire straits and the prodevelopment forces who see no need to restrain use. The vast majority of the populace, sensibly enough, holds views that are somewhere between these extremes; that balance itself is a positive sign for the future of water policy in our country. The study has made the point that we must always guard against creating unnecessary fears tending to push the public toward extreme reactions, which may, for example, lead Congress to overregulate water. In chapter 5 we saw this kind of problem with the original Clean Water Act and its amendments, and also how the requirements of the Safe Drinking Water Act may lead to vast overspending on relatively minor public health issues.

That said, this book sees no major crises at present in either water quantity or water quality. In every case examined, whether for irrigation, domestic supply, or wildlife support, there are good alternatives available to us, usually at small cost. If there are no crises, how can the vigilance and attention of the American public be maintained in the area of water policy? The answer is to build flexible institutions now that can provide leadership in the future and will be ready when serious issues requiring broad participation are again encountered. These could arise from a wide variety of causes, such as global climate change, extended drought from climate change or from natural fluctuations, acid precipitation, continued contamination of drinking water aquifers, an inability to deal effectively with non-point sources of pollution, and epidemics due to new waterborne pathogens. This book has aimed to demonstrate the overriding need to have such institutions in readiness.

Federal Water Policy: A Great Success Story?

One measure of the success of public policy is how well it has survived crises and avoided failures; from this perspective federal water policy has been remarkably successful. Over the past 200 years the evolving policies have weathered many crises, with the water resource base more or less intact and generally improved over what it was 40 years ago. The best example of its resilience is the government's response to deteriorating ambient water quality in the 1950s and 1960s. Within the remarkably short time of five years, plans and programs were put in place with the effect that by the end of the 1980s the major problems of gross water pollution from point sources were solved. But in the process another series of water quality problems were revealed—groundwater contamination, non-point sources, and man-made chemicals in lakes and river sediments. These problems (crises, some would say) remain to be resolved in the 1990s.

An important broad indicator of success for federal water policy is that per capita usage has declined over the past 25 years. In fact, despite increases in population, the United States is now using less total water than it did in 1975. The large and unexpected decline in consumption is not directly attributable to any single policy but rather to the effect of many policies. It has helped to secure the resource base and made the management of water easier.

The Need to Move Away from Crisis Management

The history of water policy assures us that current problems will be dealt with, but also that new ones will surface. The global and international issues of climate change and acid rain are often raised as the problems to be addressed in the early 2000s. One response is to take a guardedly optimistic view, like Gilbert White's, of the past and the future (White, 1983). Alternatively, one may see federal policy as lurching from crisis to crisis, offering little assurance of being able to deal with the next alarm. When faced with the severe drought of 1988–1989 in the foodgrain heartland of the country, for example, the federal government had no contingency plans in place. The White House had to scramble to put together interagency teams to find out what was taking place, and Congress had to rush through emergency relief legislation. Now that the drought is over, no agency or policy remains in place to deal with the next event. The "pollutant of the week" syndrome each time a new or unregulated chemical is discovered in trace amounts in drinking water sources is another example of the crisis mentality.

Physical and Human Infrastructure

It would be shortsighted to hold that all the water infrastructure in the United States has been completed, but it seems reasonable to say that most of the big projects requiring federal intervention have been built. The need for federal construction and development agencies is likely to decline rapidly in the future. The water resource base is secure. The United States has all the water it needs for all conceivable uses well into the middle of next century (and beyond, if the current rates of per capita decline in water use continue).

Federal water resource policies, built on our natural endowment and aided by state and local policies, have been very successful in achieving this water security. Few countries in the world are as well off in this regard as the United States at present. But if the need for future projects is greatly diminished, there is now a price to be paid, mainly by the leading federal water agencies, the Bureau of Reclamation and the Army Corps of Engineers. They face major reductions in their size and redefinitions of their missions. This means the reallocation, reassignment, and possibly retraining of large numbers of employees (maybe as many as 60,000) from these and other federal water agencies. How to accomplish this is itself a major policy question for the federal government, particularly when coupled with the downsizing of the defense and national security sector.

Regulation and Enforcement of Water Quality

The federal government is now extricating itself from the massive federal construction grant program for wastewater treatment and replacing it with a revolving loan fund. In the early 1990s the time for passing most water quality activities back to the states has come. Since the price of success in government is that you work yourself out of a job, there remains primarily the challenge of retrenching the federal water quality establishment (never nearly as large as the armies that worked in water supply). By the end of the century, federal concern with water quality should be no greater than its concern for other aspects of our natural resources and environment such as the atmosphere, land, forests, and mines.

However, lest we conclude prematurely that all construction problems have disappeared, we need to be mindful of the strange ways that regulation in one area can lead to unintended consequences in another. Section 303 of the Clean Water Act requires the establishment of total maximum daily loads (TMDLs) on water bodies where meeting the BPT or BAT emission levels of each polluter would still violate the ambient quality standards necessary for designated uses of the water body. The parts of the water bodies that fall into this category are called water quality limited segments (WQLSs). It turns out that nationwide the EPA has done little to ensure that adequate ambient water quality standards have been set (Pedersen, 1988) against which the TMDLs could be calculated, but at the same time environmental groups have been trying to force the EPA to enforce Section 303. Typically the violations of the ambient standards occur during low flow periods. The options available in such a regulatory setting are to reduce the loads allowed by each NPDES permit or to improve the ambient stream quality by augmenting the low flow. The water resource development agencies are fully aware of this choice and they are actively involved in promoting the need to build upstream storages for low flow augmentation—thus creating a new line of business for agencies faced with extinction. Of course, not only is the in-stream ambient water quality for human uses improved, but the ecosystem users also benefit. The need for the costly building of new upstream storages (or in some cases the reallocation of existing stored water to the trouble spot) will be more or less compelling depending on how important we think the in-stream values are, and on how successfully the dischargers of pollution can organize political resistance to a tightening of the standards they must meet.

Current Water Policy Initiatives

When elections loom or administrations change in Washington, there are always plenty of policy studies, commissions, and reports on all aspects of governmental policy, and water policy is no exception. The period from November 1992 through March 1993 saw at least seven major documents produced (U.S. Advisory Commission on Intergovernmental Relations, 1992; Water Quality 2000, 1992; Intergovernmental Task Force on Monitoring Water Quality, 1992; Carnegie Commission on Science, Technology, and Government, 1992; Colorado Natural Resources Law Center, 1992; Shabman, 1993; Environmental and Energy Study Institute, 1993).

As discussed in chapter 3, there is a long history of studies on how best to deal with water and environmental issues in the federal government. The most recent ones fit well in this genre. In its extremely cogent report specifically aimed at environmental research and development, the Carnegie Commission (1992) argues for major reorganizations of the federal research establishment. Most notably, it recommends moving NOAA from the Department of Commerce and merging it with the U.S. Geological Service into a new Environmental Monitoring Agency. As that document demonstrates, there are almost infinite possible combinations and permutations of agencies; the real question is whether the congressional committees would ever give up their prerogatives over the existing structure in order to allow significant change to happen.

Because of the difficulties in reassigning agencies and personnel, the U.S. Advisory Commission on Intergovernmental Relations (1992) makes recommendations that specifically avoid these kinds of action. It urges the integration of the federal pollution control laws by having Congress pass a multimedia environmental law covering discharges to air, water, and land. It couples this with a recommendation that the federal government encourage the states to administer a greater number of federal environmental standards. These proposals would go a long way toward bringing about an effective coordination of federal and state environmental work. They would not, however, address the resource development issues involved in many water situations.

Water Quality 2000's text focuses on water quality, but states that "narrowly focused water policies impede the holistic solutions that address watershed-based planning, cross-media effects, the connection between water quantity and water quality, incentives for pollution prevention, and management for environmental results." Other impediments they note are conflicts among water institutions, legislative and regulatory overlaps, inadequate funding, inadequate research and development programs, and inadequate communication with the population at large.

The Long's Peak Workshop (Colorado Natural Resources Law Center, 1992) was long on recommendations and short on time allowed to carry them out: 47 recommendations for the new Clinton administration (17 for the first 100 days and the remainder for the next four years). The recommendations cluster into four major areas: water use efficiency, ecological integrity and restoration, clean water, and equity and participation in decision making. Among the most salient of the longer-term proposals was the suggestion that "the President should create a Water Task Force composed of federal, state, and tribal governments to develop a strategy for better coordination in the development and implementation of national water policy. The Task Force should consider proposals for a new agency or other structures consolidating all federal water management functions and programs." All of the recommendations made in this book are included somewhere among the 47 Long's Peak recommendations.

Shabman (1993) finds evidence of strong acceptance by the Corps of Engineers of its "new environmental mission," which has been evolving under congressional leadership over the past decade. For example, during 1992 the Corps allocated $361 million (over 10 percent of its budget) to protection and restoration of environmental resources. He recommends that the Corps should concentrate on the watershed level of analysis and problem definition. He also sees a large and expanded role in the area of wetlands protection, restoration, and maintenance.

Chapter 3 suggests that there is very little new that can be said about water policy. Any blueprint for federal water policy should, therefore, ground itself on historical successes, and should be limited to a list of concrete actions that, if undertaken in whole or in part, will greatly improve the formulation and implementation of federal water goals. The recent studies mentioned above provide a source of inspiration for this purpose. The blueprint that follows draws upon them and the entire body of historical experiences witnessed in the earlier chapters of this book. It is divided into three parts: the first deals mainly with institutional issues; the second with specific substantive issues; the third with targets of opportunity that will present themselves in the relatively near future.

A Blueprint for Federal Water Policy: Institutions

The President's Water Council.   The value of a central federal water resources agency has been documented exhaustively in the past. Presently there is no agency within the executive branch responsible for taking a comprehensive view of federal water programs and formulating policy recommendations on that basis. At the end of 1991 the director of the Office of Management and Budget designated the U.S. Geological Survey as the lead agency for the Water Information Coordination Program (Darman, 1991) in order to ensure "effective decisionmaking for natural resources management and environmental protection at all levels of government." While this goes part of the way to addressing the need for coordination, it is a long way from what is really needed.

Specifically, there should be a water institution that is a unit of the Executive Office of the President. It would derive its principal influence from its direct relationship with the president and his or her chief administrative and policy aides, not from statutory powers and responsibilities. For that reason, an appropriate name for the new agency would be the *President's Water Council.*

The primary responsibility of such a council would be to formulate a coordinated approach to water resources nationally and to oversee the federal contribution to that effort. Among the tangible manifestations would be a timely assessment of conditions nationwide and a periodic statement of explicit objectives to be met. It would also be charged with facilitating water resource programs throughout the country. In this regard it would supplement, rather than supplant, the activities of existing federal, state, local, and private agencies; it would avoid substantial line responsibilities of its own. This would help keep the council small, competent, knowledgeable, flexible, and available.

There is no single best way to organize a function of government. Several options need exploring. The first is the interagency model the earlier Water Resources Council represented, with each of the principal federal water agencies having a seat at the policy table. The second is the independent model illustrated by the Council on Environmental Quality,[1] a three-member body appointed by the president and authorized by statute. The third option is a hybrid of the two in which certain of the line water agencies, and a number of other members (including representatives of the states), would be brought together by executive order to constitute the President's Water Council. Although the exact makeup of the council should be the prerogative of the elected administration, there are certain principles of organization that should not be overlooked.

• The council should report directly to the president through the White House domestic policy apparatus. On water matters, it would be expected to supply

programmatic advice to supplement the fiscal advice offered the president by his Office of Management and Budget.
• The council should be headed by a nonagency chairman of at least the level of a deputy or under secretary. Membership should be expanded beyond the conventional officials of cabinet rank to include representatives of the scientific and engineering communities and, especially, representatives of the state and regional levels of government. This would ensure the independent stature previous agencies have lacked.
• There must be an explicit role for the states and tribes in any federal water policy structure. Depending on the organizational option selected, representation might be from a council of advisors drawn either from the regional water councils recommended below or from the existing regional governors' conferences/associations. Even though the focus of these recommendations is on federal water policy, the recent devolution of water resources responsibilities to the states, and the states' own growing capabilities, argue strongly for a more definitive state role in establishing and carrying out federal water policy.
• A small, professional staff must be available to the President's Water Council and kept free from entangling agency allegiances and operating program burdens. The staff must be large enough to perform (or oversee) the required studies and analyses, yet small enough to prevent an independent bureaucracy from emerging with its own agenda. Extensive use of interagency memoranda of understanding and intergovernmental personnel transfers would heighten the sense of mutual objectives.

Other functions might be assigned to the council by the administration and the Congress. For example, the emerging areas of water marketing, pricing, and financing are certain to make increasing demands on any central staff. Sorting out the legal entanglements among jurisdictions, and defining distinctive roles for each of the federal, state, and local participants, could be expected to occupy substantial staff and council time. The environmental aspects of water suggest an early and collaborative relationship with the Council on Environmental Quality (or its successor agency) and the impact assessment process it oversees. And in all of its activities, the President's Water Council must view itself as performing an educational function—within government, on policymakers, and with an increasingly informed and supportive public constituency. In each instance, however, any enlarged responsibilities should be visualized as merely additive to an existing central core, the critical mass of staff necessary to perform the ongoing coordination and policy oversight functions that represent the essence of its mission.

Ultimately, of course, institutional reform in itself cannot assure the success of federal water policy coordination. The council will work only if the president and other decision makers find it useful for achieving their objectives. With the strong backing of a president who is sensitive to congressional and public attitudes, the council would stand an excellent chance of being able to influence the planning and implementation of cohesive water policies throughout the nation.

National water information program.   Employing the influence and capacities of the new federal and regional councils, there should be a fully coordinated, national water information program to gather pertinent and useful data about water at scales that are meaningful and in systems that are compatible with each other and user-friendly. A federal agency other than the council—preferably one like the U.S. Geological Survey with limited line responsibilities or mission biases—should be designated as the lead institution. Where possible, the information program should be a cooperative venture with the states and regions, and should be designed to facilitate use by state and local cooperators. It should provide for data management as well as collection. The data base should reflect jurisdictional as well as hydrological realities so that it can accommodate other important data sets (e.g., census tract information, soil and water conservation needs and inventories, and geographical information systems). Some degree of standardization in the collection, storage, and retrieval of data—federal and state—would be both prudent and cost-effective. Data common to or needed by all water programs would be centralized; specialized data would remain the province of the line agencies. However, as the National Water Commission recommended in 1973, the new President's Water Council should create and operate a modest national water data referral center simply to facilitate data retrieval and use across all water-related agencies.

The present dual system of national water summaries issued periodically by USGS and national water quality inventories prepared annually by EPA seems duplicative and confusing. At a minimum the repetition should be eliminated, and at best a single integrated report should be issued at intervals of not more than five years assessing the nation's water quantity and quality. At intervals of possibly 10 years comprehensive regional water resources assessments should be undertaken. In all cases, the goals of the assessments should be stated clearly and kept in mind throughout the process; the complexity of data collected and analyzed should be appropriate to those goals; the information should come in on time and within budget; and the

assessing agency should have enough independence to assure un-
biased analyses of federal water policies and programs. Agencies accountable to the Congress and the American public should not serve
as their own expert witnesses.

This leads to the fundamental issue of data reform, namely the
question of where in the federal establishment the water data responsibilities should be lodged. The recommendation here is for the USGS
to become the primary federal water data collection agency. This
agency already has substantial capabilities and activities of its own. It
has been awarded major coordination responsibilities by OMB. Its
cooperative program with the states is well regarded and well accepted. Further, the USGS seems to be on the threshold of an expansion of its services to the water quality field and into the expanding
area of groundwater monitoring and analysis. Before redundant and
competitive independent systems become firmly established, a three-
sided arrangement should be hammered out between USGS, EPA,
and the states. To the extent possible, all of the federal mission agencies should be encouraged to contract with and utilize the USGS's
capabilities as an alternative to supplying their own data needs.

But any agency, no matter how professional, tends to reflect its
own institutional priorities and biases. Thus it is important to distinguish between the data gathering and analysis assigned to the USGS
and the independent assessments needed to ensure comprehensive and
coordinated federal and national water resources policies. These should
become the responsibility of the recommended President's Water
Council.

National water resources research.   In keeping with the needed shift
of emphasis from water development to water management, the research effort should concentrate not only on the physical and biological sciences but increasingly on political and institutional arrangements.
The research program should include reauthorization of grants to state
water institutes, continuation of the competitive grant program available to all research institutions, and a coordinated research program
within the line water agencies. Management of the overall federal
research program should be coordinated by an entity free of line agency
responsibilities and with no competing research mission of its own,
such as the proposed President's Water Council. The council could
spend full time providing the two levels of research coordination most
needed—within and outside of the federal water research program.

Alternatively, a National Water Resources Research Center could be set up within the Department of the Interior, much as was recommended by the Council on Environmental Quality in 1984. The center would be used to articulate and direct resources toward broad categories of water research needs, establish and implement a specific research agenda, and actively monitor research to ensure its quality and relevance. The center would need to be governed (or at least advised) by an independent, broadly based board of the caliber of the Water Science and Technology Board of the National Research Council.

Water education.   A final, much-needed initiative is a concerted, public-private water education effort to make the nation more aware of its resources, problems, and needs. With support from government, the initiative should come from the private sector, employing the influence, prestige, and capacities of any number of civic, professional, and educational organizations.

There is much that can be done to make the American public aware of water realities. But the job is too large to be undertaken by a single organization or agency, even though there is a need for some entity to serve as an information switchboard and encourage the application of approaches that seem to work. The need to develop a network of educational participants seems equally urgent, so that individuals with special skills and knowledge can be more readily available to others, avoiding replication of effort.

A series of regional forums should be held around the nation to link the federal water administrators with their state, local, and private counterparts. It would be especially valuable to expose the federal policymakers firsthand to local articulation of the issues in particular parts of the country. The forums would best be sponsored by a nongovernmental organization—a civic organization like the League of Women Voters, a national research and educational entity like the Conservation Foundation, or a professional organization such as the American Water Resources Association. Important networking would be facilitated—perhaps even the system of national and regional councils called for above. One could also expect the emergence of a national agenda for water resources for the years ahead, one reflecting regional, state, and local concerns, not just the aspirations of Washington-level bureaucrats. And through the forum process, which would be fully open to the public, an important start would be made toward educating Americans on the priority issues in water resources facing particular regions and the nation as a whole.

A Blueprint for Water Policy: Broad-Based Strategic Issues

A number of important inquiries and investigations need to be launched. Because of the complex and sensitive nature of many current water issues, and the long lead time required to bring about even modest changes in policy, the following specific issues need an early examination.

Legal issues.   An early agenda item should be sorting out the legal issues relating to water in both the eastern and western parts of the country. These include conflicts between state-based legal systems, the status of reserved federal and Indian water rights, and the need to facilitate interstate compacts and other arrangements. Such examinations have been called for in countless policy studies, and remain relevant and needed today.

Water quality issues.   In water quality, there is an urgent need to review the cost effectiveness, the timetables, the attainability, and the prescriptive nature of the present technology-based standards and regulations. Integration to achieve coherence among water quality programs also merits attention, as do the rational relationships between water quality, air quality, and waste disposal standards. In particular, the neglect of ambient water quality because of excessive concern for effluent standards needs to be corrected. A legislatively authorized inquiry analogous to the 1976 National Commission on Water Quality report seems called for in this area.

Land and water issues.   The interrelationships of water and land, an explosive subject politically but a crucial one substantively, need to be addressed promptly. In the absence of a full-scale examination of this issue, land use and its impacts on water quality and quantity are being dealt with in piecemeal fashion. The role of wetlands and control of non-point sources of pollution are of particular concern in this context. An independent land and water commission, analogous to the 1973 National Water Commission, is warranted.

Drought issues. The droughts of the 1980s, though significant in their own right, also contain seeds of opportunity for longer-term water policy reforms that are at the heart of this inquiry. For example, if improved interagency coordination is one such objective, the federal drought-related apparatus instituted in 1988, and the state responses in California during 1991, could be modified to serve other needs. Networking with state and local observers for drought reporting purposes might provide the groundwork of a larger water information and extension system. The need to assemble data related to drought conditions cannot help underscoring other needs for timely and accurate information and resource assessments. The current National Study of Water Management during Drought managed by the Corps of Engineers has apparently already incorporated many of these suggestions into its study framework.

Financing issues. Finally, the most pressing puzzle of all—how the nation will finance its water program in the years ahead—deserves attention at the highest policy level. A select committee of the Congress could well be required for such purposes. When the federal debt is at historic levels, it seems incongruous that a valuable renewable resource like water is still contributing substantially to the annual budgetary deficit.

The agenda for such an inquiry should include an examination of how to provide water at the least cost to consumers and the surrounding economy, eliminate diseconomies, meet operating costs, finance infrastructure replacement, set investment priorities, and utilize modern marketing and pricing techniques to enhance water revenues. Questions of social equity in water availability should be squarely addressed, in light of the need for increased application of the beneficiary-pays principle, but in the face of the simultaneously increasing inability of many communities to pay for the improvements mandated by federal programs.

As part of this inquiry, the water development agencies should be asked to pick pilot projects in order to explore the feasibility of reallocating stored water and redeveloping existing federal projects. The establishment of a national trust fund for water should also be among the financing options examined. These recommendations reflect the strong conviction that the days of liberal federal spending for water resources projects are not likely to be repeated in the foreseeable future.

A Blueprint for Federal Water Policy: Specific Issues

Groundwater. The Department of the Interior should serve as the lead federal agency for the collection and analysis of groundwater data, and for the dissemination of information to all parties concerned. The U.S. Geological Survey would serve as the principal agency within Interior for such purposes. The 102nd Congress took no action on groundwater protection; it is important that the 103rd Congress make more progress on this critical issue, since this is the missing link in federal legislation regulating water.

Non-point source pollution. As the major unregulated contributor to ambient water degradation, non-point sources need to be adequately controlled. Non-point controls will require broad intersectoral coordination among agriculture, municipalities, and all other land uses. These issues are probably too broad for any one piece of legislation, but a start should be made by supporting the independent land and water commission mentioned above, and considering this to be part of its mandate.

No net loss of wetlands. As chapter 3 showed, the federal government has been intimately involved in the destruction of our riverine and coastal wetlands since the middle of last century. Swamps were seen as lands to be drained and brought under cultivation as quickly as possible. Recently, however, the realization has sunk in that wetlands were a valuable buffering system between terrestrial and aquatic ecosystems. With over three-quarters of our wetlands now drained and serving some other land use, it is critical that the remaining ones be protected as well as possible. The current procedures relying on Section 404 of the Clean Water Act are largely left to the Corps of Engineers to enforce (subject to an EPA veto). Since the Corps was largely responsible for aiding the states and individuals in the past to drain wetlands by levee construction programs, some view this as having the fox guard the chicken coop, despite the dramatic improvement in the Corps's approach over the past 20 years (Shabman, 1993).

However, wetlands are a land use that goes far beyond the EPA and the Corps of Engineers, and wetlands preservation necessitates the widespread participation of all federal land and water agencies along with their state and regional counterparts. This issue might be one upon which to start building the consensus needed for a President's Water Council.

Maintenance of in-stream flows.   The sustainability of aquatic ecosystems is intimately connected with the quantity and quality of water left in rivers, streams, and lakes. Damming rivers and diverting flows has had major impacts all over the United States. It is in the arid regions, however, that the effects are first noted. The fish populations of the western rivers and estuaries have been badly affected for a long time, and the recent drought in California brought the populations down to very low levels. Even rivers with plenty of water in their untamed state, such as the Columbia and Colorado, suffer badly when large amounts of water are impounded for hydropower production. The Reclamation Act of 1992 establishes the importance of maintaining adequate flows in the western rivers at all times. The extension of this principle to all other regions of the country is an important agenda item for the next few years.

Conclusions: Creating an Ongoing Self-Limiting and Self-Correcting Policy

Some readers may think this book has taken a roundabout way to reach its conclusions. Reading through so much history, hydrology, technology, economics, and political science may seem a high price to pay for such mundane conclusions. But the book followed this course to remind us that in a democracy no government policy is easy to explain or predict. Because of its technical bases, water policy may be slightly more arcane than other areas of government, but not particularly so. Serious analysts of any sphere of government policy need to explore all its foundations before they propose changes. This is what the book has been about. Because of the uncertainties inherent in managing social systems, the actual recommendations are in some sense less important than the journey to arrive at them. If we know the route well, we can adapt to any unexpected changes in the destination.

   In recommending and implementing policy it is important to create solutions that are self-limiting and self-correcting. The temptation is to create institutions that may be excellent for solving today's and yesterday's problems but are too cumbersome to be able to change to deal with tomorrow's. An example is the inability of our current institutions to deal effectively with the problems posed by non-point source pollutants. The federal water quality institutions developed piecemeal over the past 25 years were designed to address gross point source pollution. They have been remarkably successful at reducing this type of contamination in the streams, at the cost of not being adequate to deal with toxic contamination or the gross pollution from non-point sources. This is despite the provision in the relevant legislation requiring periodic renewal and amendment. Another example is the predicament in which the TVA, the Bureau of Reclamation, and to a lesser extent the Corps of Engineers currently find themselves. These are large institutions that were very successful in water resources development but have now been made largely irrelevant by their own success and the changing concerns of water policy.

Not only can problems be misperceived, but the nature of the problems themselves changes over time. For example, a marked change took place during the 1960s in the U.S. population's expectations regarding clean air and water. Social goals change; earlier, economic development was uppermost in most people's minds when water was discussed; today our society places much more emphasis on the quality of life. Changing lifestyles, political alliances, and institutions all take a toll on the ability of policy mechanisms to respond to the real issues. Other changes, such as climate change, make the task of formulating coherent and believable policies extremely difficult. Nevertheless, it is important to try.

In addition, the intellectual climate and public perceptions evolve. According to Schneider (1993) a major change of this sort is now under way. He calls it the environmental "third wave." The first wave was the mobilization of the environmental movement in the late 1960s to deal with gross water and air pollution. The second wave came in the late 1970s with the perception, fueled by some of the excesses of groundwater contamination and by greatly improved environmental monitoring, that there were many subtle chemical dangers posing great threats to human health. The third wave, according to Schneider, is the realization that we may have gone too far in regulating many chemicals in the environment beyond any economic or health justification. By spending too much on harmless or relatively harmless chemicals, we have reduced our financial and institutional capability to deal with more serious environmental challenges; the best has become the enemy of the good.

The obvious lesson that the guiding legislation should have a mandatory five-year review built into it could be bolstered by ensuring that Congress does not micromanage the legislation and the agencies responsible for regulating water. The agencies themselves should be encouraged to avoid building large staffs and programs for issues that are susceptible to rapid change.

The advantage of an institution like the proposed President's Water Council is that it would have the ability to take the long view of human needs and events. The assessment functions recommended for it are partially responsive to those concerns, but their focus is apt to be on the short or middle term. Somewhere in the new council's program should be the capacity to define long-term priorities for federally controlled water resources and regulatory programs, analyze the societal costs and benefits of setting and attaining given goals, select existing problems for extended examination and resolution, predict future crises and formulate avoidance strategies, and encourage collaborative planning by federal, state, regional, academic, nonprofit, and other nongovernmental institutions. By assuming these responsibilities as the hub of a network of existing and expanded policy study groups, and by being able to dip into many treasuries of accumulated knowledge, the council could remove policy formulation from the biased, self-interested line agency setting and set it free to give the nation a true and responsive view of its water future.

Finally, more reliance on market solutions may be the most effective route to self-limiting, self-correcting policy. Privatization of natural resources, like any other market, depends on the people's willingness to pay for them. As discussed in chapter 6, with sufficiently high tariffs many of the water infrastructure problems would disappear. Under these circumstances it would then be a natural sector to consider for private ownership, or at least private management. Many other countries, notably England and France, are using the private market for water supply and wastewater services previously provided by the government. In the United States, as the recent experiences in California and the Southwest show, there is a large and growing market for water. These markets need governmental help in regularizing property rights and creating mechanisms to ensure third-party rights against external effects.

The recommendations for federal water policy made here are largely institutional in nature. In the water area there is very little that is ruled out on technological grounds, and if we allow for desalination we can have an infinite supply of fresh water; only economics and ultimately politics will rule out specific actions. In this sense water policymakers and water managers are in a fortunate position compared to their colleagues wrestling with resource issues such as energy and social issues such as housing.

A myopic view and the broader long-term vision both recognize that water policy does not evolve on its own. It results from public debate and ongoing battles in the courts and in Congress. After that it is up to the people in the agencies and public utilities to carry out the practical day-to-day tasks of implementing the often crazy patchwork of federal water policy.

The Past as Prologue: Opening Salvos of the Clinton Administration

It is a sobering thought that in public policy analysis actual events have a habit of overtaking precise academic formulations. At the time of writing, the new Clinton administration is trying to make its own sense out of water policy issues among many other areas of interest. Already there have been some bold and dramatic moves that will change the direction federal water policy will take over the next four years.

A White House Office on Environmental Policy, replacing the Council on Environmental Quality, has been announced to coordinate environmental policy within the federal government. Its director will participate in each of the major policy councils: the National Security Council, the National Economic Council, and the Domestic Policy Council. Clearly the new administration is attuned to the needs for coordination with the federal establishment. It remains to be seen whether the areas the new office encompasses will prove too broad for one council to deal with. One could envisage that it will need at least to have subgroupings dealing with various aspects of the environment, one of which could serve effectively as the President's Water Council that is recommended in this book.

A second move has been to develop legislation to make EPA into a cabinet agency. For the reasons spelled out in this book, I would prefer that elevating EPA to cabinet status be deferred until a closer look is taken at the reorganization and merging of other resource-based and environmental agencies.

Third, the details of the president's economic stimulus package are just becoming known. In the short term, out of a proposed total package of $30 billion, $845 million will be made available for wastewater cleanup initiatives. An additional $94 million will go toward speeding up the construction of about 30 projects for flood damage reduction, inland waterway and deep-draft harbor transportation, hydropower, environmental restoration, and recreation. From 1993 until 1997 over $500 million would be spent on reducing the backlog of Corps of Engineers cyclic maintenance projects and $139 million would go to EPA's watershed resource restoration program.

On the revenue generation side the federal government is moving to exploit some of its past investments in water. Some $460 million would be raised over four years by increasing the inland waterway fuel tax from its current 19 cents per gallon to $1.19 by 1997, $65 million by improving enforcement of the harbor maintenance fees, $72 million by increasing recreation user fees at existing Corps of Engineers projects, and $398 million will be saved by reducing and stretching out construction funding for low-priority water projects. All of these measures fit in well with the recommendations of this book on how to improve the financing and maintenance of water infrastructure. They appear to be an auspicious start for the Clinton administration.

Appendix 1

Water Activities at the Federal Level

1776–1800

Minimum Government: Navigation and Navigational Aids

1783   Treaty of Independence, navigation of Mississippi River to remain free.

1784   George Washington becomes president of the Potowmack Canal Company, conflict over "illegal" interstate compact leads to Constitutional Convention.

1785   Virginia and Maryland Compact on fisheries on Chesapeake Bay and in Potomac estuary.

1786   Constitution signed, water to be a state subject.

1787   Northwest Ordinance, inland waterways "shall be common highways and forever free . . . without any tax, impost, or duty therefor."

1789   Support for navigation aids, maintenance of existing harbors, and a lighthouse on Chesapeake Bay, $3,500 appropriated.

1800–1820

Strict Constructionist: No Internal Improvements

1802   Army Corps of Engineers established at West Point.

1808   Treasury Secretary Gallatin's Report on Internal Improvements, suggests federal involvement in roads and canals.

1811   President Madison rejects federal funding for Erie Canal in message to Congress.

1820–1840

Cautious Government: The Commerce Clause

1820   Appropriations for the military include money for surveys of the Ohio and Mississippi rivers.

1821   Congress votes $150 to remove obstacles in the Thames River.

1822   Appropriations for rivers and harbors work, first such for the Army Corps of Engineers.

1824   Gibbons v. Ogden, Supreme Court supports navigation improvements, cites the "commerce clause."

1824   First appropriations for navigation works on the Ohio and Mississippi rivers subsequent to the 1820 surveys.

1824   General Survey Act, empowers president to employ Corps of Engineers to make survey plans and estimates of roads and canals of "national importance," introduces "centralized national planning."

1826   Rivers and Harbors Act, the first of many such omnibus acts giving authority to Corps of Engineers.

1828   Navigation improvement on the Tennessee River, land grant to Alabama, the proceeds of the sale to go to navigation improvement.

1840–1860

Land Hunger: The Swamp Acts

1849   An act to aid in draining the Louisiana swamplands: swamps and lands subject to riverine flooding granted to Louisiana in return for construction of levees and drains.

1850   Swamp Act, encourages draining of wetlands, extends to all states the privileges granted previously to Louisiana.

1852   Ellet Report, proposes upstream storage to control floods on the Mississippi.

1860–1880

Rapid Economic Growth: Transportation and Plans for the West

1861   Humphreys-Abbot Report, disagrees with Ellet Report and suggests that only levees are needed for flood control.

1862   Homestead Act, opens up the West to large-scale settlement.

1868   Joint resolution of Congress regarding surveys and examinations of rivers and harbors.

1871   Joint resolution of Congress regarding the preservation and protection of food fisheries, establishes the U.S. Fish and Fisheries Commission.

1873   Timber Culture Act, promotes tree planting in part to increase precipitation.

1874   Senator Windom's Select Committee on Transportation Routes to the Seaboard, supports federal river development for grain transport.

1877   Desert Land Act, sale of 640-acre tracts to persons who would irrigate them within three years, leads to massive land speculation in the West.

1879   Establishment of the Geological Survey and the Public Lands Commission.

1879   Establishment of the Mississippi River Commission.

1879   John Wesley Powell (first director of the Geological Survey) reports on arid regions of the U.S.

1880–1900

Settling the West: Private Irrigation Development

1884   Act authorizing a dam on the Mississippi River for private power development but with navigation rights reserved.

1888   Joint resolution of Congress requesting an irrigation survey by the Geological Survey.

1888   Appropriation for Geological Survey includes funds for the study of public lands for irrigable areas and withdrawal of them from sale.

1890   Repeal of Geological Survey's authority to reserve irrigable lands.

1891   Payson Act, the president authorized to reserve public forest lands from sale to protect future water supplies.

1893   Quarantine Act to prevent the interstate spread of contagious diseases.

1894   Carey Irrigation Act, grants public lands to states for irrigation purposes, limited to 160-acre tracts.

1899   Rivers and harbors appropriations establish federal standards for bridges over navigable streams.

1899   Regulation of development at mineral springs, authorizes permits for private development.

1899   Rivers and Harbors Act (Refuse Act), prohibits the discharge of solid materials into navigable streams.

1900–1920

Progressive Movement: Conservationism

1902   Reclamation Act (Newlands Act), establishes the Reclamation Fund from sale of public lands in 16 western states, the fund to be used for irrigation development, and establishes the Reclamation Service, which later becomes the Bureau of Reclamation.

1902   Board of Engineers for Rivers and Harbors created in the Corps of Engineers to review all reports, goal to eliminate unsound projects.

1905   Forest Service created in the Department of Agriculture, Gifford Pinchot appointed Chief Forester, he introduces policy of limiting permits for water power to 50 years and charging fees, eliminates congressional giveaways.

1906   Reclamation Act of 1906, authorizes the sale of surplus power from reclamation projects.

1906   General Dam Act, regulates private dam construction on navigable streams.

1906   Antiquities Act, prohibits the destruction of historic or prehistoric ruins or monuments on public land.

1908   Report of the Inland Waterways Commission, recommends comprehensive planning and coordinated development of water resources and a "National Waterways Commission."

1909   Report of the National Conservation Commission, also advocates comprehensive planning and development of water resources.

1910   Amendment to the General Dam Act of 1906, places further restrictions on dams on navigable streams.

1910   Withdrawal Act, president authorized to withdraw public lands from sale for irrigation, water power, and other purposes.

1910   Rivers and harbors appropriations, Corps of Engineers authorized to consider power after navigation.

1911   Weeks Act, authorizes the purchase of forest land to preserve streamflow.

1912   Report of the National Waterways Commission, recommends comprehensive and integrated water development.

1912   Power production authorized at navigation dams.

1912   Public Health Service Act, establishes the Public Health Service and authorizes it to study stream pollution.

1912   Winters Doctrine enunciated by the Supreme Court, federal government assumed to have reserved prior rights to water for lands that it reserved for Indian tribes and other federal uses.

1913   Raker Act, San Francisco allowed to build Hetch Hetchy Dam in Yosemite National Park.

1914   Public Health Service promulgates drinking water standards for interstate carriers.

1916   Establishment of the National Park Service.

1916   Appropriations for fishery research.

1917   Waterways Commission set up for national water resources planning, never meets because of war.

1917   Flood control works on the Mississippi and Sacramento rivers authorized, the first federal construction for flood control.

1917   River Regulation amendment, authorizes the Waterways Commission to coordinate water resources development.

1920–1930

Normalcy: Conflicts over Public Power

1920   Federal Water Power Act, empowers the already existing Federal Power Commission, until then a cabinet-level committee, to regulate nonfederal power and to sell power from federal projects, also abolishes the Waterways Commission.

1921   President Harding halts work on the Wilson Dam power project on the Tennessee River.

1922   Colorado River Compact, apportions the Colorado River waters between Arizona, California, Colorado, Nevada, New Mexico, Utah, and Wyoming.

1924   Funds appropriated to purchase private lands to establish the Upper Mississippi River Wildlife and Fish Refuge.

1924   Oil Pollution Act, prohibits the discharge of oil into coastal navigable waters.

1924   Fact Finders Act, requires the Secretary of the Interior to obtain detailed information about reclamation projects before submitting them to Congress.

1925   Rivers and Harbors Act, the Corps of Engineers authorized to survey all navigable rivers, except the Colorado, and formulate general plans covering irrigation, navigation, power production, and flood control (the so-called 308 plans took more than 20 years to complete).

1928   An act for the control of floods on the Mississippi, first steps away from the levees-only policy of the Humphreys-Abbot Report.

1928   Boulder Canyon Dam (Hoover Dam) authorized, with electric power as a paying partner.

1930–1940

The New Deal: Power, Irrigation, and Flood Control

1930   Federal Power Commission reorganized as an independent agency.

1933   The New Deal, establishment of the Public Works Administration, many sewage treatment plants are financed.

1933   Tennessee Valley Authority Act, "a corporation clothed with the power of government but possessed of the flexibility and initiative of a private enterprise" (FDR).

1934   Creation of National Resources Board, stresses coordination of water resources development, later becomes the National Resources Planning Board.

1934   Fish and Wildlife Coordination Act, provisions to be made in federal projects for wildlife if economically practical.

1934   Migratory Bird Hunting Stamp Act, provides revenue source for acquiring refuges.

1935   Works Progress Administration established.

1936   Flood Control Act of 1936, the first nationwide flood control act, makes an upstream-downstream division between the Department of Agriculture and the Corps of Engineers; Section I requires benefit-cost analysis for the first time for federal water projects.

1937   Water Facilities Act, authorizes the Department of Agriculture to provide irrigation facilities.

1937   Congress rejects a proposal by President Roosevelt to create seven regional resource planning agencies like TVA; the Chief of Engineers, Major General Schley, testifies against the proposal.

1938   Flood Control Act of 1938, authorizes 100 percent federal funding of flood control works.

1939   Reorganization Act, specifically enjoins the president from transferring away any functions of the Corps of Engineers.

1939   Reclamation Project Act, authorizes the Secretary of Interior to plan and construct multiple-purpose projects.

1939   Tripartite agreement between War, Interior, and Agriculture to provide voluntary interagency consultation in river basin surveys.

1940–1950

War and Peace: Water Supply for Economic Development

1943   Congress does away with the National Resources Planning Board (NRPB) and blocks its functions from being picked up by successor agencies.

1943   Executive Order 9384 directs all public works construction agencies to prepare and update long-range programs that must be submitted annually to the Bureau of the Budget, an attempt to retain the power of the NRPB.

1943   Federal Inter-Agency River Basin Committee (FIARBC) formed from old tripartite group.

1944   Pick-Sloan plan for joint development of the Missouri River by the Corps of Engineers and the Bureau of Reclamation.

1944   Flood Control Act of 1944, recognizes irrigation, power, and recreation as secondary purposes of federal flood control projects; Southwestern and Southeastern Power Administrations created to market the power.

1944   Joint resolution concerning the Fish and Wildlife Service authorizing a survey of marine and freshwater fishery resources.

1945   Flood Control Act of 1945, power facilities to be provided at flood control reservoirs.

1946   President Truman issues a memorandum on power policy that coordinates federal power development under the Department of the Interior.

1946   Legislative Reorganization Act, consolidates numerous standing committees of both houses of Congress and, with minor changes in 1961, still dictates the disposition of water bills.

1948   Water Pollution Control Act, provides for construction loans and technical assistance to municipalities (no construction money is ever appropriated), Public Health Service abatement procedures for interstate waters but no enforcement without state consent.

1949   Report of the first Hoover Commission on the Reorganization of the Executive Branch of the Government, recommends consolidating all federal water functions under the Department of the Interior, also an independent review board under the Office of the President.

1950–1960

Rapid Economic Growth: Concern about the Resource Base

1950   Report of the Water Resources Policy Commission (Cooke Commission), favors a Department of Natural Resources and river basin commissions.

1950   Dingell-Johnson Act, authorizes federal excise tax redistributed to states for fish restoration and management.

1950   FIARBC releases the "Green Book" on "proposed practices for economic analysis of river basin projects," recommended for use by all federal agencies.

1951   Engineers Joint Council Report.

1952   Bureau of the Budget Circular A-47, standards for accepting or rejecting agency projects.

1952   Report of the Materials Policy Commission (Paley Commission).

1952   Saline Water Act, funding for research on desalination.

1952   Jones Subcommittee (House) to Study Civil Works, proposes co-ordination through congressional policy determination and project authorization.

1953   Report of the Missouri River Basin Survey Commission.

1954   President Eisenhower requests FIARBC be reconstituted as the Inter Agency Committee on Water Resources (IACWR or "icewater"), with subcabinet-rank members.

1954   Small Watershed Act, establishes the Soil Conservation Service's small watershed program.

1954   Flood Control Act of 1954, expands the recreational use of flood control reservoirs.

1954   Bureau of the Budget revises Circular A-47 to incorporate partnership principle for flood control costs.

1955   Report of second Hoover Commission's Task Force on Water Resources and Power, recommends strengthening the Bureau of the Budget to evaluate projects and creation of a Water Resources Board in the Executive Office of the President.

1955   Report of cabinet-level President's Advisory Committee on Water Resources Policy, recommends more executive-level control through a board of review and a permanent IACWR, introduces the importance of environment and water quality.

1956   Small Reclamation Projects Act, authorizes the Secretary of the Interior to provide local organizations with loans to construct and rehabilitate small irrigation projects.

1956   Upper Colorado Storage Project authorized, reservoirs banned in national parks and irrigation limited to basic agricultural commodities.

1956   Fish and Wildlife Act, establishes the Fish and Wildlife Service in the Department of the Interior.

1956   Federal Water Pollution Control Act, extends and strengthens the 1948 act and adds $50 million per year as direct grants toward construction of sewage treatment plants; enforcement procedures strengthened.

1956   Senate Resolution 281, negative response to the President's Advisory Committee report, also opposes Bureau of the Budget Circular A-47 and its proposed revision.

1958   Congress establishes two "United States Study Commissions" to draw up comprehensive river basin plans, one study in Texas, the other in the Southeast.

1958   Fish and Wildlife Coordination Act, wildlife preservation to receive equal consideration in water resources projects.

1958   Federal Boating Act, federal jurisdiction over boating regulations.

1958   National Outdoor Recreation Resources Review Commission established.

1958   Water Supply Act, makes future urban and industrial water supplies an equal purpose in the planning of multiple-purpose projects.

1959   President Eisenhower vetoes the Public Works Appropriation Act for fiscal 1960.

1960–1970

The Affluent Society: Concern for Recreation and Aesthetics

1961   Report of the Senate Select Committee on National Water Resources (Kerr Committee), predicts a water crisis mainly for municipal and industrial water users, highlights pollution as a major future cost, urges comprehensive river basin planning.

1961   President Kennedy responds with a draft of the Water Resources Planning Act including a proposed Water Resources Council with cabinet-level participation.

1961   Delaware River Basin Compact, an interstate compact (Delaware, New Jersey, New York, and Pennsylvania) establishing the Delaware River Basin Commission with broad powers, concern for municipal water supply and quality, and for recreation.

1961   Amendments to the 1956 Federal Water Pollution Control Act, a further $100 million per year in grants for construction of sewage treatment plants authorized.

1962   Senate Document 97 on policies, standards, and criteria, advocates a Water Resources Council; President Kennedy adopts it and rescinds Circular A-47.

1962   Wetlands drainage limited, an attempt to resolve conflicts between federal agencies.

1962   Food and Agriculture Act, allows for cost sharing in Soil Conservation Service's small watershed program.

1962   Rivers and Harbors Act, reiterates and expands authority for recreation development in Corps of Engineers projects.

1962   National Academy of Sciences report (Abel Wolman), sees urgent need for education in all water disciplines.

1963   Bureau of Outdoor Recreation established.

1963   Federal Council for Science and Technology report, virtually identical conclusions as the NAS report.

1964   Water Resources Research Act, water resources research institutes established in each state and $100,000 grants for research made to each.

1964   Wilderness Act, preservation of certain areas of national parks from any development.

1964   Land and Water Conservation Fund Act, financing of outdoor recreation by admission stickers.

1965   Water Resources Planning Act, establishes the Water Resources Council and authorizes river basin commissions, bans the study of interbasin transfers.

1965   Water Project Recreation Act, recreation recognized as a legitimate purpose in multipurpose reservoir projects.

1965   Anadromous and Great Lakes Fish Act, promotes preservation and development of fish resources.

1965   Water Quality Act establishes Federal Water Pollution Control Administration (FWPCA) in HEW, sets timetable for states to create water quality standards for their interstate waters, and increases construction grant program to $150 million per year.

1966   Johnson's message to Congress cites pollution as the major natural resource problem.

1966   FWPCA transferred to the Department of the Interior.

1966   Endangered Species Act passed.

1966   Clean Waters Restoration Act, radically increases program funding in increments to $1.25 billion per year by fiscal 1971, and increases federal participation from 30 to 50 percent of construction costs.

1968   Estuarine Protection Act, authorizes the study of estuaries to examine the desirability of establishing a nationwide system of estuarine areas.

1968   Congressional White Paper on a National Policy for the Environment, reflects the hearings on 20 bills filed during the 90th Congress on national environmental policy.

1969   Executive Order 11472 establishes a Cabinet Committee on the Environment to advise the president on environmental quality matters (terminated by Executive Order 11541 in 1970).

1970–1980

Unbridled Environmentalism: Ecology and Quality

1970   Water Quality Improvement Act, regulation of oil spills after Santa Barbara oil spill.

1970   National Environmental Policy Act of 1969 (NEPA) signed into law, environmental impact statements for any significant federal action, sets up the Council on Environmental Quality (CEQ).

1970   President Nixon establishes the Environmental Protection Agency (EPA) to consolidate federal environmental control programs.

1972   Clean Water Act, $18 billion in grants paying up to 75 percent of cost for municipal sewage treatment plants and interceptor sewers, pivotal water pollution legislation.

1973   Report of the National Water Commission.

1974   Safe Drinking Water Act, first comprehensive federal drinking water legislation, emphasizes monitoring, training, and demonstration, only small amount of grant money.

1976   Report of the National Commission on Water Quality.

1977   President Carter attempts to cancel a list of water projects.

1977   Clean Water Act amendments, more grant money.

1978   Message from the president on federal water policy initiatives.

1978   Water Resources Council second assessment of the nation's water resources.

1978   Waterways user charge/Lock and Dam 26, act passed authorizing an Inland Waterways and Harbor Maintenance Trust Fund to be financed by a tax on diesel used by commercial water transport.

1979   Water Resources Council, "Principles and Standards," promulgates uniform standards for all federal water projects.

1980–1992

The New Federalism: Devolution of Powers and Funding

1980   Interior Department report on water policy implementation.

1981   Clean Water Act amendments, reduce federal share of construction funds to 50 percent, encourage innovative solutions.

1981   Water Resources Council zero-budgeted.

1982   Cabinet-level Council on Natural Resources created.

1982   Office of Water Policy created in the Department of the Interior.

1982   Reclamation Reform Act, revises the rules concerning upper limits on subsidized acreage.

1982   Circular A-67 of the OMB, directs data acquisition of some 28 federal agencies.

1983   Scandal at the EPA, Administrator Gorsuch resigns.

1986   Safe Drinking Water Act amendments, heavy emphasis on regulation, promulgation of standards for 83 contaminants, requires filtration for most surface sources.

1986   Water Resources Development Act, authorizes $16.5 billion in new water resources projects, cost sharing of up to 100 percent for some projects, a major watershed in federal water policy.

1986   Office of Management and Budget, "Principles and Guidelines."

1987   Clean Water Act amendments, further reduce federal grant money, cost sharing to be phased out over five years and replaced by revolving loan funds.

1988   Water Resources Development Act.

1990   Water Resources Development Act.

1992   Water Resources Development Act.

1992   Reclamation Projects Authorization and Adjustment Act.

1992   Energy Policy Act.

Appendix 2

Significant Policy Studies of the Twentieth Century

1. Report of the Inland Waterways Commission, February 1908. Senate Document 325, 60th Congress, 1st Session.

2. Joint Conservation Conference of National and State Officials, 1908.

3. White House Conference of Governors, 1908.

4. Report of the National Conservation Commission, January 1909. Senate Document 676, 60th Congress, 2nd Session.

5. North American Conservation Conference, 1909.

6. Report of the National Waterways Commission, 1912.

7. Report of the President's Committee on Water Flow, 1934. House Document 395, 73rd Congress, 2nd Session.

8. Reports of the National Resources Board and successor agencies, 1935–1944.

9. U.S. Water Resources Committee, 1936. Drainage basin problems and programs.

10. Report of the U.S. Commission on Organization of the Federal Government (first Hoover Commission), June 30, 1950. Water resources and power.

11. Report of the President's Water Resources Policy Commission (Cooke Commission), 1950. A water policy for the American people. Volume 1.

12. Report of the President's Materials Policy Commission (Paley Commission), June 2, 1952. Resources for freedom.

13. House Committee on Public Works (Jones Subcommittee), December 5, 1952. Water resources policy. Committee Print Nos. 21, 22, 23, and 24. 82nd Congress, 2nd Session.

14. Report of the President's Advisory Committee on Water Resources Policy, 1956. House Document 315, 84th Congress, 2nd Session.

15. Report of the Senate Select Committee on National Water Resources (Kerr Committee), January 30, 1961. Senate Document 29, 87th Congress, 1st Session.

16. Report of the National Water Commission, June 14, 1973. Water policies for the future.

17. Report of the National Commission on Water Quality, 1976.

18. Hearings before the Subcommittee on Water Resources, Senate Committee on Environment and Public Works, March 31–April 18, 1977.

19. Message from the President of the United States, June 6, 1978. Federal water policy initiatives.

20. U.S. Water Resources Council, 1978. The nation's water resources, 1975–2000. 4 volumes.

21. Office of Water Policy, Department of the Interior, 1983. Water in America, 1983.

Based upon Black, 1987, pp. 92–93.

Appendix 3

Water History: A Nonfederal Perspective

1650–1750

Private Local Solutions

1652	Boston	Private Water-Works Company chartered, uses wooden conduits.
1682	Philadelphia	First two public water supply wells sunk.
1704	Boston	Francis Thrasker builds sewer with private capital.
1707	New Orleans	First levees built.
1709	Boston	Act regulating drains, common shores, and charges for the use of drains.
1741	New York	Law requiring city to install pumps at the expense of the users.

1750–1800

Public Investment for Citywide Water and Sewers

1754	Bethlehem, Pa.	First municipal waterworks built by Hans Christopher Christiansen.
1769	New York	City builds trunk sewer lines.
1774	New York	Christopher Colles builds a public waterworks and citywide distribution system.
1784	Virginia	George Washington becomes president of the Potowmack Canal Company.
1795	Boston	The Aqueduct Corporation is authorized to pipe water four miles from Roxbury.

1800–1830
Water Power, Canal Building, and Development of Distant Sources

1801	Philadelphia	Major waterworks and distribution system built by Benjamin Latrobe, including two large steam pumps.
1804	Paisley, Scotland	First public water supply sand and gravel filtering system.
1817	Boston	40 miles of pine pipe supply 800 families for an annual fee of $10 ($1,000 at current rates).
1817–19	Cincinnati	Cincinnati Manufacturing Company given a 99-year monopoly, horse-driven pumps.
1817–25	New York	Governor De Witt Clinton builds the 364-mile Erie Canal from Albany to Buffalo.
1818	Philadelphia	Bored wood pipes replaced by cast iron for the first time in the U.S.
1820	Lowell, Mass.	Industrial water power developed for textile industry.
1823	Boston	City repeals 1709 act and assumes public control over private sewers.
1827	Chelsea, England	James Simpson designs and builds the prototype slow sand filter.
1827	Greenock, Scotland	Robert Thom installs reverse flow filters (still used in modern plants).

1830–1870
Massive Urban Epidemics of Cholera and Typhoid

1832	New York	Cholera epidemic attributed to polluted city wells claims 3,500 lives; Croton Aqueduct built in response, completed in 1842 at a cost of $13 million.
1832	Richmond	First attempt to filter public water supplies fails (not to be successfully implemented until 1872).

1837	Philadelphia	3.1 mgd supplied to 19,600 water users (158 gallons per capita per day).
1840s		Softening using lime and soda ash discovered.
1842	Chicago	Chicago Hydraulic Company given a 70-year charter to provide water.
1843	Hamburg, Germany	First city with comprehensive sewer system.
1844	Boston	Due to excessive wastewater production, ordinance prohibits baths without doctor's orders.
1847	Utah	Mormon irrigation begins; within 15 years 150,000 acres are under irrigation.
1848	Chicago	Severe typhoid epidemic; private water company replaced by municipal utility in 1851.
1848	Boston	$4 million 14-mile Cochichuate Aqueduct completed.
1850	Massachusetts	Lemuel Shattuck presents the first State Sanitary Report.
1850		Francis turbine invented to boost efficiency of water power.
1857	New York	Julius W. Adams uses Hamburg as his model for citywide sewers.
1857		Louis Pasteur announces the germ theory of disease.
1857	Anaheim, Calif.	German farmers develop several thousand acres of irrigated land on the Santa Ana River.
1858	Chicago	Citywide sewers.
1862	New York	A billion-gallon reservoir completed in Central Park.
1869	Massachusetts	State Board of Health inaugurates first serious investigation of waste treatment methods.
1869	St. Louis	James P. Kirkwood's Report on the Filtration of River Waters.

1870–1900

Private Irrigation and Electric Power, Large Investments in Urban Water Infrastructure

1870	Greeley, Colo.	The Greeley Colony established on 32,000 acres of irrigated land.
1872	Augusta, Maine	First sewage farm in U.S. opened at the State Insane Asylum.
1874	New York	Filtered water supply.
1876	Missouri	James P. Kirkwood's report to the State Health Board giving evidence that streams cleanse themselves.
1880s	Lawrence, Mass.	Intermittent filtration for sewage treatment.
1880	Memphis	First separate sewers.
1882	Appleton, Wisc.	First hydroelectric plant (12.5 kw).
1887	Massachusetts	Lawrence Experimental Station opened.
1887	New York	Chemical precipitation for sewage treatment.
1887	Kansas City	Chemical precipitation in water purification using alum.
1889	Chicago	Contact beds introduced for the first time.
1889	Boston	MIT initiates sanitary engineering program.
1890	Massachusetts	Lawrence Laboratory Report on sewage treatment (Hazen, Fuller, and Sedgwick).
1895	Niagara Falls	11,200 kw of power generated.
1896	Louisville	Experiments with chlorine disinfection.

1900–1920

Chlorination of Drinking Water and Secondary Sewage Treatment

1900		Ozone employed for water disinfection in Europe.
1902	Middelkerke, Belgium	First permanent chlorination of water supply.
1902	Little Falls, N.J.	First municipal rapid sand filter (plant is still in operation).

1903	Boston	MIT opens pollution research laboratory.
1904		Copper sulphate discovered to be an effective algicide, used for control of tastes and odors.
1905	Oberlin, Ohio	First municipal water softening plant.
1908	Jersey City	First continuous U.S. application of chlorine.
1908	Reading, Pa.	First trickling filter.
1909	Fort Meyer	Liquid chlorine (compressed chlorine gas) becomes commercially available.
1911	Madison, N.J.	Imhoff tanks introduced from Germany.
1912		Streeter and Phelps introduce the concept of biochemical oxygen demand (BOD).
1914	Milwaukee	Large-scale research on activated sludge, leading to world's largest sewage treatment plant by 1924.
1916	San Marcos, Texas	First activated sludge plant installed.
1916	Chicago	Chlorine added at lake pumping stations.
1917	Chicago	Chemical conditioning of sludge.

1920–1970

"Modern" Water Supply and Wastewater Treatment

1920s		Activated carbon used for taste and odor control.
1920s		Oxidation ponds and stabilization lagoons promoted.
1921	Milwaukee	Vacuum filtration of sludge.
1932	Chicago	Sludge incinerator installed.
1945		Introduction of fluoride to reduce tooth decay.
1965	Boston	Deer Island treatment plant opened.

1970–1990

Widespread Secondary Sewage Treatment and Concern for Toxics in Drinking Water

1984	Boston	Judge Paul Garrity threatens injunction on Boston Harbor.
1985	Boston	Formation of Massachusetts Water Resource Authority.
1987	Boston	Work started to bring Boston and surrounding towns into compliance with 1972 federal water laws.

Appendix 4

Federal Institutions with a Role in Formulating or Implementing Water Policy

Department of Agriculture

Agricultural Stabilization and Conservation Service (1961)
Economic Research Service
Soil Conservation Service (1935)
U.S. Forest Service (1905)
Farmers Home Administration (1949)

Department of Commerce

Economic Development Administration (1965)
National Oceanic and Atmospheric Administration (NOAA) (1970)
National Weather Service

Department of Defense

Corps of Engineers (1779)
Board of Engineers of Rivers and Harbors (1902)

Department of Energy

Federal Energy Regulatory Commission (1930)
Bonneville Power Administration (1937)
Southwestern Power Administration (1943)
Southeastern Power Administration (1950)
Alaska Power Administration (1976)
Western Area Power Administration (1977)

Department of Housing and Urban Development

Community Planning and Development (1974)

Department of the Interior

Bureau of Indian Affairs (1849)
Bureau of Land Management (1946)
Bureau of Mines (1910)
Office of Surface Mining Reclamation and Enforcement (1977)
Bureau of Reclamation (1902)
Bureau of Outdoor Recreation (1963)
Fish and Wildlife Service (1871)
Geological Survey (USGS) (1879)
National Park Service (1916)

Department of Justice

Land and Natural Resources Division

Department of Labor

Occupational Safety and Health Administration (1970)

Department of State

Oceans, International Environment, and Scientific Affairs Office

Department of Transportation

Maritime Administration
U.S. Coast Guard (1915)
St. Lawrence Seaway Development Corporation (1954)
Office of Pipeline Safety (1977)
Office of Hazardous Materials Transport (1977)

Executive Office of the President

Council on Environmental Quality (1970)
Office of Management and Budget (OMB) (1970)
Office of Science and Technology Policy (1970)
Office of Policy Development (1981)

U.S. Congress

Committees and subcommittees (see appendix 5)
Congressional Budget Office
Congressional Research Bureau
Office of Technology Assessment (1972)

Independent agencies and government corporations

Environmental Protection Agency (EPA) (1970)
Federal Emergency Management Agency (1978)
Interstate Commerce Commission (1887)
General Accounting Office
Tennessee Valley Authority (TVA) (1933)
Appalachian Regional Commission (1965)
Federal Maritime Commission (1961)
Panama Canal Commission (1979)

Boards, committees, and commissions

Delaware River Basin Commission
Susquehanna River Basin Commission
Mississippi River Commission
Pacific Northwest Electric Power and Conservation Planning Council

Bilateral organizations

International Joint Commission (U.S.-Canada) (1911)
International Boundary and Water Commission (U.S.-Mexico) (1889)

Federal courts

Appendix 5

Standing Congressional Committees with Jurisdiction over Water (102nd Congress)

House of Representatives

Committee	Subcommittees
Agriculture	Conservation, Credit, and Rural Development
	Cotton, Rice, and Sugar
	Department Operations, Research, and Foreign Agriculture
	Domestic Marketing and Consumer Relations, and Nutrition
	Forests, Family Farms, and Energy
	Livestock, Dairy, and Poultry
	Peanuts and Tobacco
	Wheat, Soybeans, and Feed Grains
Appropriations	Commerce, Justice, State, Judiciary, and Related Agencies
	Defense
	District of Columbia
	Energy and Water Development
	Foreign Operations, Export Financing, and Related Programs
	Interior and Related Agencies
	Labor, Health and Human Services, Education and Related Agencies
	Legislative
	Military Construction
	Transportation

	Treasury, Postal Service, and General Government
	VA, HUD, and Independent Agencies
Armed Services	Defense Policy Panel
	Department of Energy Defense Nuclear Facilities Panel
	Environmental Restoration Panel
	Future Uses of Defense Manufacturing and Technology Resources Panel
	Investigations
	Military Education Panel
	Military Installation and Facilities
	Military Personnel and Compensation
	Morale, Welfare, and Recreation Panel
	North Atlantic Assembly Panel
	Procurement and Military Nuclear Systems
	Readiness
	Research and Development
	Seapower and Strategic and Critical Materials
	Structure of U.S. Defense Industrial Base Panel
Budget	Budget Process, Reconciliation, and Enforcement
	Community Development and Natural Resources
	Defense, Foreign Policy, and Space
	Economic Policy, Projections, and Revenues
	Human Resources
	Urgent Fiscal Issues
Education and Labor	Elementary, Secondary, and Vocational Education
	Employment Opportunities
	Health and Safety
	Human Resources
	Labor-Management Relations

	Labor Standards
	Postsecondary Education
Select Education	
Energy and Commerce	Commerce, Consumer Protection, and Competitiveness
	Energy and Power
	Health and the Environment
	Oversight and Investigations
	Telecommunications and Finance
	Transportation and Hazardous Materials
Government Operations	Commerce, Consumer, and Monetary Affairs
	Employment and Housing
	Environment, Energy, and Natural Resources
	Government Information, Justice, and Agriculture
	Human Resources and Intergovernmental Relations
	Legislature and National Security
Interior and Insular Affairs	Energy and Environment
	General Oversight and California Desert Lands
	Insular and International Affairs
	Mining and Natural Resources
	National Parks and Public Lands
	Water, Power, and Offshore Energy Resources
Judiciary	Administrative Law and Government Relations
	Civil and Constitutional Rights
	Crime and Criminal Justice
	Economic and Commercial Law
	Intellectual Property and Judicial Administration
	International Law, Immigration, and Refugees

Merchant Marine and Fisheries	Coast Guard and Navigation
	Fisheries and Wildlife Conservation, and the Environment
	Merchant Marine
	Oceanography, Great Lakes, and the Outer Continental Shelf
	Oversight and Investigations
Public Works and Transportation	Aviation
	Economic Development
	Investigations and Oversight
	Public Buildings and Grounds
	Surface Transportation
	Water Resources
Science, Space, and Technology	Energy
	Environment
	Investigations and Oversight
	Science
	Space
	Technology and Competitiveness
Small Business	Antitrust, Impact of Degradation, and Ecology
	Environment and Employment
	Exports, Tax Policy, and Special Problems
	Procurement, Tourism, and Rural Development
	Regulation, Business Opportunities, and Energy
	SBA, the General Economy, and Minority Enterprise Development
Ways and Means	Health
	Human Resources
	Oversight
	Select Revenue Measures
	Social Security
	Trade

Senate

Committee	Subcommittees
Agriculture, Nutrition, and Forestry	Agricultural Credit
	Agricultural Production and Stabilization of Prices
	Agricultural Research and General Legislation
	Conservation and Forestry
	Domestic and Foreign Marketing and Product Promotion
	Nutrition and Investigations
	Rural Development and Rural Electrification
Appropriations	Agriculture, Rural Development, and Related Agencies
	Commerce, Justice, State, Judiciary, and Related Agencies
	Defense
	District of Columbia
	Energy and Water Development
	Foreign Operations
	Interior and Related Agencies
	Labor, Health and Human Services, Education, and Related Agencies
	Legislative Branch
	Military Construction
	Transportation
	Treasury, Postal Service, and General Government
	VA, HUD, and Independent Agencies
Banking, Housing, and Urban Affairs	Consumer and Regulatory Affairs
	Housing and Urban Affairs
	International Finance and Monetary Policy
	Securities
Budget	No subcommittees

Commerce, Science, and Transportation	Aviation
	Communications
	Consumer Affairs
	Foreign Commerce and Tourism
	Merchant Marine
	National Ocean Policy Study
	Science, Technology, and Space
	Surface Transportation
Energy and Natural Resources	Energy Regulation and Conservation
	Energy Research and Development
	Mineral Resources Development and Production
	Public Lands, National Parks, and Forests
	Water and Power
Environment and Public Works	Environmental Protection
	Gulf Pollution Task Force
	Nuclear Regulation
	Superfund, Ocean and Water Protection
	Toxic Substances, Research and Development
	Water Resources, Transportation, and Infrastructure
Finance	Deficits, Debt Management, and International Debt
	Energy and Agricultural Taxation
	Health for Families and the Uninsured
	International Trade
	Medicare and Long-Term Care
	Private Retirement Plans and Oversight of the IRS
	Social Security and Family Policy
	Taxation
Foreign Relations	African Affairs
	East Asian and Pacific Affairs
	European Affairs
	International Economic Policy, Trade, Oceans, and Environment

	Near Eastern and South Asian Affairs
	Terrorism, Narcotics, and International Operations
	Western Hemisphere and Peace Corps
Governmental Affairs	Federal Services, Post Office, and Civil Service
	General Services, Federalism, and the District of Columbia
	Government Information and Regulation
	Oversight of Government Management
	Permanent Subcommittee on Investigations
Judiciary	Antitrust, Monopolies, and Business Rights
	Constitution
	Courts and Administrative Practice
	Immigration and Refugee Affairs
	Juvenile Justice
	Patents, Copyrights, and Trademarks
	Technology and the Law
Labor and Human Resources	Aging
	Children, Family, Drugs, and Alcoholism
	Disability Policy
	Education, Arts, and Humanities
	Employment and Productivity
	Labor
Small Business	Competitiveness and Economic Opportunity
	Expert Expansion
	Government Contracting and Paperwork Reductions
	Innovation, Technology, and Productivity
	Rural Economy and Family Farming
	Urban and Minority-Owned Business Development

Appendix 6

Chronology of Waterway User Charge Legislation

1787   Northwest Ordinance, calls inland waterways "common highways and forever free . . . without any tax, impost, or duty therefor."

1938   President Franklin D. Roosevelt orders a study on taxation of waterborne commerce. From then until 1978 every presidential commission recommends a user charge, and every president from Roosevelt to Gerald Ford calls for a waterway user tax.

1973   January. Freshman Senator Pete V. Domenici (R, New Mexico) is assigned to the Senate Public Works Committee.

1974   Sometime during the work of the committee, staffer Hal Brayman, who had been promoting a waterway user charge for several years without success, explains barge industry's "free ride" to Domenici, who decides to pass a waterway user charge.

1976   Domenici formulates a "hostage plan" to link a user charge to the authorization bill for Lock and Dam 26. During late summer, after two weeks of hearings on Lock and Dam 26, Senator Domenici asks committee staffers to draft a bill on user charges. Over the next few weeks they draft a bill, helped by a variety of railroad lobbyists, public interest, and environmental groups. The charges are designed to cover 50 percent of the cost of waterway improvements.

1977   February 24, Senator Domenici files his Lock and Dam 26/User Charge Bill, designated S.790. Jointly assigned by the parliamentarian to Commerce and to Public Works.

February 26(?), H.R.4339 filed by Berkley Bedell (D, Iowa). Assigned jointly to Public Works and Ways and Means.

March 4, a "request" bill for a waterway user charge, originally sent to Congress on January 19 by the outgoing Ford administration, is assigned to Commerce.

April 1, Water Resources Subcommittee starts hearings on S.790. 50 witnesses requested to testify.

May 2, Brock Adams, Secretary of Transportation, testifies, "it is no longer in the national interest to continue direct taxpayer support of commercial water transportation." He also proposes to collect the user charge by means of a tax of 42 cents per gallon on diesel fuel for barges. Administration also against Lock and Dam 26; Adams is for the hostage plan.

May 3, mark-up of the bill with no mention of a fuel tax, with about $350 million per year in revenue from tolls and fees.

May 5, mark-up by full Public Works Committee, which votes 14 to 1 to report the bill to the Senate floor.

May 12, Senator Magnuson, Commerce's chairman, agrees to waive hearings and go directly to a full committee mark-up. Senator Long's version drafted by barge lobbyists, requiring a delay in any action on user charges for 18 months pending a further study, is adopted.

May 15, deadline for bills with expenditures to get into the budget planning.

May 20, Adams sends a letter to the Democratic and Republican leadership of the Senate indicating "the President's very firm intention to veto any bill authorizing construction of a new Lock and Dam 26 . . . which does not also contain a provision for the establishment of a waterway user charge."

June 19, Senator Howard Baker (R, Tennessee) announces he will vote for Domenici's bill.

June 22, Floor debate on user charge/lock and dam bill starts at 10 a.m. with five senators present. Senator Stevenson (D, Illinois) offers an amendment that the authorization proceed but that the user charge be held for an 18-month study. At 3:58 p.m. a vote is called on the amendment, which is defeated 51 to 44. S.790 comes for a vote at 7:00 p.m. and the vote is 71 to 20 in favor. Senator Mike Gravel (D, Alaska), chairman of the Water Resources Subcommittee, as previously arranged with Domenici, proposes that the bill just passed be inserted into H.R.5885, the omnibus rivers and harbors bill recently passed in the House. This is approved without objection.

June 24, barge lobbyists raise the origination clause with Representative Al Ullman (D, Oregon), chairman of House Ways and Means Committee. (His district is heavily involved with water transport on the Columbia River.)

Late June, Ullman formally asks that H.R.5885 be held at the Speaker's desk and requests that a "blue slip" be prepared.

July 15, Speaker O'Neill arranges compromise between Ullman and Johnson, whereby H.R.5885 will be allowed to die and the House will pass its own version of S.790, with Johnson's Public Works Committee approving a new lock and dam authorization bill and Ullman's Ways and Means Committee writing a waterway tax bill, both to be reported out of the committees within two weeks.

July 18, Water Resources Subcommittee holds hearings on waterway toll bill, and is urged by barge interests to have fuel tax not exceeding 4 cents per gallon. Adams testifies and asks for 40 cents per gallon.

July 19, Water Resources Committee marks up and reports out a waterway bill as agreed between Ullman, Johnson, and representatives of barge industry.

July 20, Public Works Committee holds its mark-up and approves the subcommittee bill.

July 21, Ways and Means begins hearings on a fuel tax plan.

July (Monday following hearings which ended on a Friday), Ways and Means meets to mark up bill. There is some unhappiness by railroad interests at a 4 cent tax. Finally agreed to have it 4 cents for the first two years, going up to 6 cents thereafter. Ullman directs the committee staff to prepare a joint committee report with the Public Works Committee.

July 29, in two weeks, as agreed, the two committees have reported out a waterway bill. But O'Neill decides that it should not be presented to the House until after the farm bill.

September, Ullman asks Rules Committee for a "modified open rule" (allowing floor amendments to reduce the tax, not allowing those to increase it). Bedell objects and a "closed rule" (yes or no on the fuel tax) is adopted. Debate on the rule is scheduled for October 6 and on the bill itself the following Tuesday, to be completed on Thursday, October 13.

October 6, closed rule approved.

October 13, waterway bill H.R.8309 passes 331 to 70 and is sent to the Senate. Jennings Randolph, Chairman of the Senate Public Works Committee, picks H.R.8309 as the omnibus legislation for waterway pork barrel projects.

October, Domenici succeeds in having H.R.8309 held on the Senate floor without referral to committees. Long calls a hearing on the "general subject of waterway taxes."

November, Senator Robert C. Byrd approaches both Long and Domenici to see if a vote could be scheduled to see whether the Senate would like the tough user charges of the now lapsed S.790, the milder tax approach of H.R.8309, or no charge at all. No agreement is reached.

1978 By April H.R.8309 has at least 60 amendments, each authorizing a project that is important for at least one senator.

Domenici breaks off negotiations with the barge interests and works a deal with Senator Adlai Stevenson to have a fuel tax gradually increasing from 4 cents to 12 cents per gallon and also a capital recovery feature recovering a fixed percentage of the government expenditures each year.

May 3, the Stevenson-Domenici bill, as an amendment to H.R.8309, is defeated 43 to 47, Long's amendment (only the 4 to 12 cent tax) is passed 88 to 2, and H.R.8309 now contains more than 90 amendments authorizing $2 billion in pork barrel projects. Later in the day Jimmy Carter says he will veto the bill. O'Neill announces that he will delay appointment of House conferees until the House's own omnibus bill is ready.

June, House Public Works Committee holds hearings on its own omnibus bill. The committee ends up with about $2 billion in projects. Carter warns O'Neill that he will veto the legislation, particularly if it amounts to $4 billion. O'Neill refuses to appoint House conferees for the Senate version of H.R.8309.

During the summer Long offers to negotiate. Shapiro (Long's unofficial chief negotiator) and Brayman decide to start all over again with a clean bill, to get away from the $4 billion of authorizations now likely with H.R.8309 and also to deal with the capital recovery issue. By mid-August they arrive at the solution of a flat 10 cent tax with the revenues going into an "Inland Waterways Trust Fund." (Congressional appropriations are still allowed to exceed the actual trust fund.)

Mid-September, Domenici, Long, and Adams get together and agree to the clean bill solution.

Mid-September, Adlai Stevenson slips through a new bill authorizing Lock and Dam 26 without the user charge as an amendment to pending legislation. Domenici and Long both furious.

October 6, a broad cross section of the participants agree to the new clean bill. However, O'Neill has promised the House an October 14 adjournment.

October 10, Senator Long amends an already passed tax bill, H.R.8533 dealing with tax exemption for charitable bingo games, as a vehicle for the waterways tax. Unfortunately the Senate has inadvertently substituted the waterways package for the bingo tax, not added it.

Three days of negotiations ensue with O'Neill; it is agreed that the House will pass H.R.8533 in the form the Senate approved it, and both houses will pass concurrent resolutions to put the bingo tax exemption back in.

October 13, one year, seven months, three weeks, and one day after it was introduced, the House of Representatives votes 287 to 123 in favor of H.R.8533, with Berkley Bedell leading the opposition.

October 28, President Jimmy Carter signs the bill into law.

1979   January 15, the 96th Congress opens and Congressman Bedell files a bill entitled "Transportation Users Equity Act of 1979" proposing waterway user charges.

Notes

1 Water Resources and Public Policy

1. The categories are: roads, streets, bridges, and highways; mass transit; airports; intermodal transportation; water resources; water supply; wastewater treatment; solid waste; and hazardous waste.

2. Wetland area has fallen from 215 million acres in the 1700s to 95 million acres in the 1980s.

2 Basic Hydrology, Water Demand, and Setting Standards

1. Stable runoff is that part of runoff that comes from underground flow into rivers—sometimes called the "base flow" or the sustained or fair-weather runoff. This is a conservative estimate of available water since it ignores seasonal diversion and storage possibilities.

2. See Pedersen (1988) and chapter 8 for more discussion of this point.

3. A paper by James et al., (1969) explored the relative impacts of uncertainty in data on the choices implied in a river basin setting and concluded that the outcomes were much more sensitive to socioeconomic parameters, such as future demands for water, than to the uncertain physical and biological data. This is a theme reiterated by Martin, Ingram, and Griffin (1984) in their discussion of the municipal water supply system in Tucson, Arizona (see chapter 6 for a further discussion of the Tucson case).

4. See the National Academy of Sciences' landmark five-volume study, *Drinking Water and Health* (Assembly of Life Sciences, 1977–1983), for more details.

3 History of Water Policy in the United States

1. By Admiral Ben Moreell, the head of the Task Force on Water Resources of the Second Hoover Commission (1955), Theodore Schad, the executive director of the National Water Commission (1988), and Henry Caulfield, a former director of the Water Resources Council and, since 1939, a major participant in all major water policy developments (1988).

2. Appendix 1 summarizes the development of water policy, and appendix 2, based upon Black (1987), lists the significant water policy studies of this century.

3. For more detailed treatment of these events, the reader is directed to the works by Rosen (1958) and the American Public Works Association (1976).

4 Water as a Resource: What Makes It Different?

1. For example, none of the dire predictions made in 1972 by the Club of Rome study have come to pass—the United States managed to increase its GNP by 30 percent with no increase in total energy use from 1973 to 1985—and while population and resource issues are still very difficult global problems there has been no system collapse, nor is one seriously considered imminent.

2. This was later changed by the 1987 amendments to the Clean Water Act except for industries.

3. Some specific acts are:

• Coastal Zone Management Act of 1972, as amended: PL 92-583; 16 USC 1451 et seq.
• Wild and Scenic Rivers Act of 1968, as amended: PL 90-542; 16 USC 1271 et seq.
• Coastal Resources Barrier Act of 1982: PL 97-348; 16 USC 3501–3510.
• Fish and Wildlife Coordination Act of 1958, as amended; PL 85-624; 16 USC 661.
• Marine Mammal Protection Act of 1972, as amended; PL 92-552; 16 USC 1361.
• Marine Protection, Research, and Sanctuaries Act of 1972; PL 92-532; 33 USC 1401.
• Reservoir Salvage Act of 1960; PL 86-523; 16 USC 469.

4. Other pieces of legislation that fall into the cross compliance category are:

• Endangered Species Act of 1973, PL 93-205; 16 USC 1531.
• Archeological Resources Protection Act of 1979; PL 96-95; 16 USC 470.
• National Historic Preservation Act of 1966, as amended; PL 89-665; 16 USC 470.
• Archeological Data Conservation Act of 1974; PL 93-291; 16 USC 469.
• Native American Religious Freedom Act of 1978; Pl 95-341; 42 USC 1996.

6 Economic, Financial, and Public Expenditure Imperatives

1. The creation of property rights by government water pollution regulation is discussed in chapter 4.

2. For example, Howe and Linaweaver (1967) found a price elasticity of −1.57 for residential outdoor use in 11 eastern U.S. systems and only −0.23 for indoor use in the same systems. Commercial and industrial production of wastewater is also price responsive. Elliott (1973) found that for industrial water use a 1 percent increase in water charges reduced wastewater effluents by 0.75 percent (a price

elasticity of −0.75). He also reported that increasing the charge for wastewater effluents from industry reduced the demand for water supplied to the industry with an elasticity of −0.44. Miglino and Harrington (1984), studying industrial water use in São Paulo, Brazil, found that the introduction of a separate wastewater tariff led to a 30 percent drop in water demanded and a resulting 37 percent shortfall of revenue for the water utility.

3. Billings and Day (1983) showed that the Tucson Water Department estimated that the water rates would have to be increased by 17.6 percent in order to pay for the increasing cost of the city supply from the Central Arizona Project, but the authors pointed out that that would be too low by a factor of three when the price elasticity is taken into account. Billings and Day estimated that a rate increase of 59 percent would be necessary to increase the total revenue by 17.6 percent.

4. Recent data reported by the *Water Strategist* during July 1989, however, shows a marked softening of the water markets.

7 Political Imperatives: Legislative, Executive, and Bureaucratic

1. See Harleman (1990) for a completely contradictory view on the Boston case.

2. In the Senate alone, barge operators could count on the support of James O. Eastland (D, Miss.), Chairman of the Judiciary Committee, John C. Stennis (D, Miss.), Chairman of the Armed Services Committee, Russell B. Long (D, La.), Chairman of the Finance Committee, and John L. McClellan (D, Ark.), Chairman of the Appropriations Committee, all from waterway states bordering the Mississippi River, and, from Washington, the only important western waterway state, Warren Magnuson (D), Chairman of the Commerce Committee, and Henry (Scoop) Jackson (D), Chairman of the Energy Committee.

3. The conservation organizations included the Izaak Walton League, the Environmental Policy Center, and the Coalition for American Rivers; the good government interest was represented by the Public Interest Economic Center and the National Taxpayers Association.

8 Institutional Needs and Possible Responses

1. "Institutions" is used here in the broadest sense: Congress is itself an institution, as are the Corps of Engineers and the American Chemical Manufacturers Association. "Institutional needs" refers to activities as different as the mechanism whereby one executive agency coordinates its programs with another or the need to create entirely new agencies.

Much of this chapter was presented earlier in a paper by Foster and Rogers (1988).

2. The act allocated $18 billion in expenditures over a five-year period, with $9 billion going toward direct subsidies and the other $9 billion being given to the states as seed money for "State Revolving Loan Funds." The revolving funds are meant to be used in perpetuity by the states to lend money to localities to con-

struct wastewater treatment facilities, using the repayment of the loans to pay for other communities' projects. As the authorization deadline approaches there is great concern in Congress to extend the date beyond 1994.

3. By the National Research Council (1981), the General Accounting Office (1981), and the President's Council on Environmental Quality (1984).

4. By February 1993 the U.S. Geological Survey's research grant program had been zeroed out of the 1993 budget.

5. The Corps of Engineers has already carried out some studies on the benefits of potential reallocation of storage (U.S. Army Corps of Engineers, 1989). But further questions might be raised about the continued wisdom of the dry bed reservoirs maintained by the Corps of Engineers and Soil Conservation Service for flood control; the nontransferability of reclamation project water rights; and the nonrecovery of even part of the average 17 mil/kwh differential between the market value of power from conventional sources and the usual rates for federal hydropower. Project redevelopment in this sense, of course, can in many cases create a revenue source large enough to meet needs and recover costs, and also to fund future water initiatives.

6. It is likely that some form of groundwater protection will be included in the expected reauthorization of the Clean Water Act during the 103rd Congress.

7. The earlier attempts at controlling non-point source pollution were associated with Section 208 of the 1972 Clean Water Act (PL 92-500), the areawide waste treatment planning assigned to the Environmental Protection Agency in 1975. More recently, the amendments mandated by Section 319 of PL 99-100 require the states to complete non-point source assessment reports by April 4, 1988, and have full management plans (including implementation schedules) in effect four years thereafter.

In January 1985 a diverse group of scientists, bureaucrats, and state and local officials, organized as a special Federal/State/Local Non-point Source Task Force, underscored the current state of policy uncertainty by describing the limited role it felt the federal government should play in combating non-point source pollution. In its judgment, the federal agencies should furnish financial and technical assistance, coordinate interagency and state actions, assist with program development, and promote the provision of incentives to achieve water quality goals where needed.

Title XII of the Food Security Act of 1985 mandated formal conservation plans for each of the nation's 1.5 million farm ownerships by 1990, a goal that was not achieved, and calls for implementing those plans by 1995.

9 A Blueprint for Water Policy

1. The Clinton administration announced on February 8, 1993, that the Council on Environmental Quality had been replaced by the new White House Council on Environmental Policy.

References

Abelson, Philip H. 1993. "Regulatory Costs." Editorial. *Science* 259 (January 8), p. 159.

Agricola, G. 1950 (1920). *De re metallica.* Trans. Herbert C. Hoover and Lou Henry Hoover. New York: Dover.

American Public Works Association. 1976. *History of Public Works in the United States 1776–1976.* Chicago: American Public Works Association.

American Water Works Association. 1972. *Water Rates.* AWWA Manual M1. 2d ed. Denver: American Water Works Association.

Ames, Bruce N., Renae Magaw, and Lois Swirsky Gold. 1987. "Ranking Possible Carcinogenic Hazards." *Science* 236 (April), pp. 271–280.

Anderson, Terry. 1983a. *Water Crisis: Ending the Policy Drought.* Baltimore: Johns Hopkins University Press.

Anderson, Terry L., ed. 1983b. *Water Rights: Scarce Resource Allocation, Bureaucracy, and the Environment.* Cambridge, Mass.: Ballinger.

Assembly of Life Sciences, Safe Drinking Water Committee. 1977–1983. *Drinking Water and Health.* 5 vols. Washington: National Academy of Sciences.

Barnett, Harold J., and Chandler Morse. 1963. *Scarcity and Growth.* Baltimore: Johns Hopkins University Press.

Barney, Gerald O. 1981. *The Global 2000 Report to the President.* Washington: Council on Environmental Quality and the Department of State, Superintendent of Documents.

Becker, G. S., B. A. Cody, and R. M. Chite. 1991. "The Californian Drought: Effects on Agriculture and Related Resources." *CRS Report to Congress.* Washington: Congressional Research Service, Library of Congress. February 25.

Billings, R. B., and W. M. Day. 1983. "Elasticity of Demand for Water: Policy Implications for Southern Arizona." *Arizona Review* 3, pp. 1–11.

Black, Peter. 1987. *Conservation of Water and Related Land Resources.* Totowa, N.J.: Rowman and Littlefield.

Born, Stephen M., ed. 1989. *Redefining National Water Policy: New Roles and Directions*. Special Publication 89-1. American Water Resources Association, Bethesda, Md.

Calhoun, John C. 1853. *The Works of John C. Calhoun*. 6 vols. New York: D. Appleton and Co.

Carnegie Commission on Science, Technology, and Government. 1992. *Environmental Research and Development: Strengthening the Federal Infrastructure*. December.

Carson, Rachel. 1962. *Silent Spring*. New York: Fawcett Crest.

Caulfield, Henry P. 1988. "The Politics of Federal Water Policy." Seminar, Kennedy School of Government, Harvard University. February 24.

Ciriacy-Wantrup, S. V. 1963. *Resource Conservation: Economics and Policies*. Berkeley: University of California, Division of Agricultural Sciences.

Clean Water Council. 1990. *America's Environmental Infrastructure: A Water and Wastewater Investment Study*. Washington: Clean Water Council.

Coase, R. 1960. "The Problem of Social Cost." *Journal of Law and Economics* 3 (October), pp. 1–44.

Colorado Natural Resources Law Center. 1992. *America's Waters: A New Era of Sustainability*. Report of the Long's Peak Working Group on National Water Policy. Boulder: University of Colorado. December.

Congressional Budget Office. 1983. *Efficient Investments in Water Resources: Issues and Options*. Washington: Congress of the United States.

Congressional Research Service. 1980. *State and National Water Use Trends to the Year 2000*. Report no. 96-12. Washington: Congressional Research Service, Library of Congress.

Cortner, H. J., and J. Auburg. 1988. "Water Resources Policy: Old Models and New Realities." *Water Resources Bulletin* (American Water Resources Association) 24, no. 5 (October), pp. 1049–1056.

Crouch, E. A. C., R. Wilson, and L. Zeise. 1983. "The Risks of Drinking Water." *Water Resources Research* 19, no. 6 (December), pp. 1359–1375.

Darman, Richard. 1991. "Memorandum on the Coordination of Water Resources Information." Office of Management and Budget. December 10.

Dasgupta, P., S. A. Marglin, and A. Sen. 1972. *Guidelines for Project Evaluation*. New York: United Nations Industrial Development Organization.

Desvouges, W. H., and V. Kerry Smith. 1983. *Benefit-Cost Assessment for Water Programs*. Vol. 1. Prepared for the U.S. Environmental Protection Agency. Research Triangle Institute, North Carolina.

Dorfman, R., and N. Dorfman. 1972. *Economics of the Environment*. New York: W. W. Norton.

Downing, D., and S. Sessions. 1985. "Innovative Water Quality-Based Permitting: A Policy Perspective." *Journal of the Water Pollution Control Federation* 57, no. 5.

Eckstein, O. 1958. *Water Resource Development and the Economics of Project Evaluation*. Cambridge: Harvard University Press.

Elliott, R. D. 1973. "Economic Study of the Effect of Municipal Sewer Surcharges on Industrial Wastes and Water Usage." *Water Resources Research* 9, no. 5 (October), pp. 1121–1131.

Environmental and Energy Study Institute. 1993. *1993 Briefing Book on Environmental and Energy Legislation*. Environmental and Energy Study Institute, Washington.

Falkenmark, Malin. 1989. "The Massive Water Scarcity Now Facing Africa—Why Isn't It Being Addressed?" *Ambio* 18, no. 2.

Falkenmark, M., M. Garn, and R. Cestti. 1990. "Water Resources: A Call for New Ways of Thinking." INUWS Working Paper. World Bank. March.

Firestone, David B., and Frank C. Reed. 1983. *Environmental Law for Non-Lawyers*. Ann Arbor: Ann Arbor Science.

Fisher, A., and R. Raucher. 1984. "Intrinsic Benefits of Improved Water Quality." In V. Kerry Smith, ed., *Advances in Applied Micro-Economics,* vol. 3. Greenwich, Conn.: JAI Press.

Foster, Charles H. W., and Peter P. Rogers. 1988. "Federal Water Policy: Toward an Agenda for Action." Discussion Paper E-88-05. Energy and Environmental Policy Center, Kennedy School of Government, Harvard University.

Frazer, James G. 1890. *The Golden Bough: A Study in Comparative Religion*. London: Macmillan.

Freeman, A. Myrick. 1982. *Air and Water Pollution Control. A Benefit-Cost Assessment*. New York: John Wiley.

Garrett, Wilbur E. 1987. "George Washington's Potowmack Canal." *National Geographic Magazine* (June), pp. 716–753.

General Accounting Office. 1977. *Improvements Needed by the Water Resources Council and the River Basin Commissions to Achieve the Objectives of the Water Resources Planning Act of 1965*. CED-77-1. Report of the Comptroller General of the United States. Washington: U.S. Government Printing Office. October 31.

General Accounting Office. 1981. *Federal-Interstate Compact Commissions: Useful Mechanisms for Planning and Managing River Basin Operations*. CED-81-34. Report of the Comptroller General of the United States. Washington: U.S. Government Printing Office. February 20.

Getches, David H. 1990 (1984). *Water Law in a Nutshell*. 2d ed. St. Paul: West Publishing Co.

Gleick, P. H., and L. Nash. 1991. "The Societal and Environmental Costs of the Continuing California Drought." Berkeley: Pacific Institute for Studies in Development, Environment, and Security. July.

Hanke, S. H., and R. Davis. 1973. "Potential for Marginal Cost Pricing in Water Resources Management." *Water Resources Research* 9, no. 4 (August).

Harleman, Donald R. F. 1990. "Cutting the Waste in Wastewater Cleanups." *Technology Review*, pp. 59–68.

Heclo, Hugh. 1977. *A Government of Strangers: Executive Politics in Washington.* Washington: Brookings Institution.

Herfindahl, Orris, and Allen V. Kneese. 1974. *Economic Theory of Natural Resources.* Resources for the Future, Inc. Columbus: Charles E. Merrill Publishing Co.

Hirshleifer, J., J. C. De Haven, and J. W. Milliman. 1960. *Water Supply: Economics, Technology, and Policy.* Chicago: University of Chicago Press.

Hodges, Laurent. 1973. *Environmental Pollution.* New York: Holt, Rinehart and Winston.

Holmes, B. H. 1972. *A History of Federal Water Resources Programs, 1800–1960.* PB-295 733. Washington: U.S. Department of Agriculture, Economic Research Service.

Holmes, Beatrice H. 1979. *History of Water Resources Programs and Policies, 1961–1970.* Miscellaneous Publication no. 1379. Washington: U.S. Department of Agriculture, Economics, Statistics, and Cooperative Service.

Howe, Charles W. 1979. *Natural Resource Economics: Issues, Analysis, and Policy.* New York: John Wiley.

Howe, C. W., and F. P. Linaweaver, Jr. 1967. "The Impact of Price on Residential Water Demand and Its Relation to System Design and Price Structure." *Water Resources Research* 3, no. 1, pp. 13–32.

Hudson, J. F. 1981. *Pollution Pricing: Industrial Responses to Wastewater Charges.* Lexington, Mass.: Lexington Books.

Intergovernmental Task Force on Monitoring Water Quality. 1992. *Ambient Water-Quality Monitoring in the United States.* Washington: U.S. Geological Survey. December.

Interstate Conference on Water Policy. 1990. *Toward National Water Policy: The Challenge of Improving Intergovernmental Relations.* Washington: Interstate Conference on Water Policy. February.

James, I. C., B. T. Bower, and N. C. Matalas. 1969. "Relative Importance of Variables in Water Resources Planning." *Water Resources Research* 5, no. 6 (December), pp. 1165–1173.

James, L. Douglas, and Robert R. Lee. 1971. *Economics of Water Resources Planning.* New York: McGraw-Hill.

Kennedy, David N. 1991. "Allocating California's Water Supplies during the Current Drought." Keynote speech presented at the Water Resources Management Workshop, World Bank, Washington, June 24–28.

Kneese, Allen V. 1964. *The Economics of Regional Water Quality Management*. Baltimore: Johns Hopkins University Press.

Kneese, A. V. 1984. *Measuring the Benefits of Clean Air and Water*. Washington: Resources for the Future.

Little, M. D., and J. A. Mirrlees. 1974. *Project Appraisal and Planning for Developing Countries*. New York: Basic Books.

L'vovich, M. I. 1979. *World Water Resources and Their Future*. Trans. by the American Geophysical Society. Ed. Raymond L. Nace. Washington: American Geophysical Union.

Maass, Arthur. 1951. *Muddy Waters: The Army Engineers and the Nation's Rivers*. Cambridge: Harvard University Press.

Maass, Arthur. 1968. "Conservation: Political and Social Aspects." In David L. Sills, ed., *International Encyclopedia of the Social Sciences,* vol. 3, pp. 271–279. New York: Macmillan.

Maass, A., et al. 1962. *The Design of Water-Resources Systems: New Techniques for Relating Economic Objectives, Engineering Analysis, and Government Planning*. Cambridge: Harvard University Press.

Maloney, M. T., and Bruce Yandle. 1983. "Building Markets for Tradeable Pollution Rights." Chapter 9 in Anderson, 1983b.

Mann, P. C. 1981. *Water Service: Regulation and Rate Reform*. Columbus: National Regulatory Research Institute.

Martin, William E., H. M. Ingram, and A. H. Griffin. 1984. *Saving Water in a Desert City*. Washington: Resources for the Future.

Meadows, Donella, et al. 1972. *Limits to Growth*. New York: Universe Books.

Meier, G. M. 1983. *Pricing Policy for Development Management*. Baltimore: Published for the World Bank by Johns Hopkins University Press.

Meta Systems, Inc. 1985. *A Methodological Approach to an Economic Analysis of the Beneficial Outcomes of Water Quality Improvement from Sewage Treatment and Combined Sewer Overflow Controls*. Washington: Environmental Protection Agency, Office of Policy Analysis.

Metcalf and Eddy, Inc. 1991. *Wastewater Engineering: Treatment, Disposal, and Reuse*. 3d ed. New York: McGraw-Hill.

Miglino, L. C. P., and J. J. Harrington. 1984. "O Impacto da Tarifa na Geração de Efluentes Industriais." *Revista DAE* 44, no. 138 (September), pp. 212–220.

Mishan, E. J. 1976. *Cost-Benefit Analysis*. New York: Praeger.

Moreell, Ben. 1955. *Report on Water Resources and Power*. 3 vols. Prepared for the Commission on Organization of the Executive Branch of the Government by the Task Force on Water Resources and Power. June.

Mugler, Mark W. 1982. *The Federal Interest: Case Studies of Congressional Objectives and Rationale for Federal Activity*. Policy Study 82-0600. U.S. Army Corps of Engineers, Engineer Institute for Water Resources.

National Council on Public Works Improvement. 1988. *Fragile Foundations: Final Report to the President and Congress*. February.

National Groundwater Policy Forum. 1987. *Groundwater: Saving the Unseen Resource*. Washington: Conservation Foundation.

National Research Council. 1981. *The 5-Year Outlook on Science and Technology, 1981: Source Materials*. Washington: National Science Foundation.

National Water Commission. 1973. *Water Policies for the Future*. Washington: U.S. Government Printing Office.

Office of Management and Budget. 1978. "Staff Analysis: Natural Resources Reorganization." Washington: Executive Office of the President. December 16.

Office of Management and Budget. 1992. *Budget of the United States Government FY 1992*. Washington: Superintendent of Government Documents.

Passell, Peter. 1989. "Life's Risks: Balancing Fear against Reality of Statistics." *New York Times,* May 8, p. A1.

Pastor, Robert. 1980. *Congress and the Politics of U.S. Foreign Economic Policy*. Berkeley and Los Angeles: University of California Press.

Peabody, N. S., ed. 1991. "Water Policy Innovations in California." Water Resource and Irrigation Policy Program, Case Study no. 1. Arlington, Va.: Winrock International Institute for Agricultural Development. July.

Pedersen, William F. 1988. "Turning the Tide on Water Quality." *Ecology Law Quarterly* 15, p. 69.

Plater, Zygmunt. 1982 "Reflected in a River: Agency Accountability and the TVA Tellico Dam Case." *Tennessee Law Review* 49, no. 4 (Summer).

Porter, Roger B. 1980. *Presidential Decision Making: The Economic Policy Board*. New York: Cambridge University Press.

President's Council on Environmental Quality. 1984. *Environmental Quality*. 15th Annual Report. Washington.

Project 88. 1991. *Incentives for Action: Designing Market-Based Environmental Strategies*. A Public Policy Study Sponsored by Senators Timothy E. Wirth and John Heinz. Washington. May.

Reid, T. R. 1980. *Congressional Odyssey: The Saga of a Senate Bill*. San Francisco: W. H. Freeman.

Reisner, Marc. 1986. *Cadillac Desert: The American West and Its Disappearing Water.* New York: Viking Penguin.

Reisner, Marc, and Sarah Bates. 1990. *Overtapped Oasis: Reform or Revolution for Western Water.* Washington: Island Press.

Repetto, Robert, ed. 1985. *The Global Possible: Resources, Development and the New Century.* New Haven: Yale University Press.

Reuss, Martin, and Paul K. Walker. 1983. *Financing Water Resources Development: A Brief History.* EP 870-1-13. U.S. Army Corps of Engineers, Office of the Chief. July.

Rosen, George. 1958. *History of Public Health.* New York: MD Publications.

Saliba, B. C., and D. B. Bush. 1987. *Water Markets in Theory and Practice.* Boulder: Westview Press.

Sander, William. 1985. "The State of Economic Analysis in Water Resources Planning." *American Journal of Economics and Sociology* 44, no. 1 (January), pp. 121–128.

Schad, Theodore. 1988. "Evolution of a National Water Policy." Paper presented at the 15th Annual Conference of the Water Resources Planning and Management Division, American Society of Civil Engineers, Norfolk, Va. June 1.

Schneider, Keith. 1993. "New View Calls Environmental Policy Misguided." *New York Times,* March 21, p. A1.

Scott, Anthony. 1955. *Natural Resources: The Economics of Conservation.* Toronto: University of Toronto Press.

Shabman, Leonard. 1993. *Environmental Activities in Corps of Engineers Water Resources Programs.* Fort Belvoir, Va.: U.S. Army Corps of Engineers, Institute for Water Resources. January.

Sheer, D. P. 1986. "Managing Water Supplies to Increase Water Availability." Pp. 101–112 of *National Water Summary 1985.* Water Supply Paper 2300. Washington: U.S. Geological Survey.

Slovic, Paul. 1987. "Perception of Risk." *Science* 236 (April 17), pp. 280–285.

Smil, Vaclav. 1985. *Carbon, Nitrogen, Sulphur: Human Interference in Grand Biospheric Cycles.* New York: Plenum.

Solley, Wayne B., Robert R. Pierce, and Howard A. Perlman. 1993. "Estimated Use of Water in the United States, 1990." U.S. Geological Survey Circular 1081. Washington: U.S. Geological Survey.

Starr, Chauncey. 1969. "What Is Our Society Willing to Pay for Safety: Social Benefits versus Technological Risk." *Science* 165 (September 19), p. 1232.

Stavins, R., and Z. Willey. 1983. "Trading Water Conservation Investments for Water." In R. J. Charbeneau, ed., *Regional and State Water Resources Planning and Management.* Bethesda, Md.: American Water Resources Association.

Stroup, Richard L., and John A. Baden. 1983. *Natural Resources: Bureaucratic Myths and Environmental Management*. San Francisco: Pacific Institute for Public Policy Research.

Tabors, Richard D., Michael H. Shapiro, and Peter P. Rogers. 1976. *Land Use and the Pipe*. Lexington, Mass.: Lexington Books.

Tchobanoglous, G., and E. D. Schroeder. 1985. *Water Quality*. Reading, Mass.: Addison-Wesley.

UNIDO. 1978. *Guide to Practical Project Appraisal*. Vienna: United Nations Industrial Development Organization.

U.S. Advisory Commission on Intergovernmental Relations. 1992. *Intergovernmental Decisionmaking for Environmental Protection and Public Works*. Report A-122. U.S. Advisory Commission on Intergovernmental Relations, Washington. November.

U.S. Army Corps of Engineers. 1989. "Water Supply Reallocation Request: Richard B. Russell Reservoir, Elberton, Georgia." Savannah District, Savannah, Georgia. July 20.

U.S. Army Corps of Engineers. 1990. *Vision 21: A Strategic Assessment of the Nation's Water Resources Needs*. Washington: Directorate of Civil Works, U.S. Army Corps of Engineers.

U.S. Army Corps of Engineers. 1991. *The National Study of Water Management during Drought: Report on the First Year*. IWR Report 91-nds-1. Fort Belvoir, Va.: Institute for Water Resources. May.

U.S. Congress. 1961. Senate. Report of the Select Committee on National Water Resources. Report no. 29. 87th Congress, 1st session. Washington: U.S. Government Printing Office. January.

U.S. Department of Commerce. 1989. *Statistical Abstract of the United States: 1989*. Washington: U.S. Government Printing Office.

U.S. Department of Housing and Urban Development. 1984. *Residential Water Conservation Projects*. Summary Report. June.

U.S. Department of the Interior. 1987. *Bureau of Reclamation Assessment '87: A New Direction for the Bureau of Reclamation*. Washington.

U.S. Environmental Protection Agency. 1987. *Unfinished Business: A Comparative Assessment of Environmental Problems*. Washington: Office of Policy Analysis, Environmental Protection Agency. February.

U.S. Geological Survey. 1986. *National Water Summary 1985: Hydrologic Events and Surface Water Resources*. Water Supply Paper 2300. Washington: U.S. Government Printing Office.

U.S. Water Resources Council. 1977. "Nationwide and Regional Analyses for 1975 Assessment." Unpublished.

U.S. Water Resources Council. 1978. *The Nation's Water Resources 1975–2000*. Washington: U.S. Government Printing Office. December.

U.S. Water Resources Council. 1983. *Principles and Guidelines for Water and Related Land Resources Planning*. Washington: Water Resources Council.

Water Quality 2000. 1992. *A National Water Agenda for the 21st Century*. Alexandria, Va.: Water Pollution Control Federation/Water Quality 2000. November.

Waterstone, Marvin, and William B. Lord. 1989. "How Safe Is Safe?" *Phi Kappa Phi Journal* (Winter), pp. 22–25.

Water Strategist. 1989. Vol. 3, no. 2 (July).

Welsh, F. 1985. *How to Create a Water Crisis*. Boulder: Johnson Books.

Werick, William J. 1993. "A National Study of Water Management during Drought: Results Oriented Water Resources Management." Paper presented at the 20th Anniversary Conference of the Water Resources Planning and Management Division of the American Society of Civil Engineers, Washington. May 3–5.

Western Governors Association. 1989. "White Paper on Federal Water Policy Coordination." Western Governors Association, Denver, May 11.

White, Gilbert F. 1971. *Strategies of American Water Management*. Ann Arbor: Ann Arbor Paperbacks.

White, Gilbert F. 1983. "Water Resource Adequacy: Illusion and Reality." *Natural Resources Forum*, United Nations, New York.

Wollman, N., and G. E. Bonem. 1971. *The Outlook for Water-Quality, Quantity and National Growth*. Resources for the Future Inc. Baltimore: Johns Hopkins University Press.

Woodman, Jocelyn H. 1989. "Why Not Zero Waste?" *EPA Journal,* November/December, pp. 39–40.

Index